ENERGY SCIENCE, ENGINEERING AND TECHNOLOGY

SOLAR POWER, RENEWABLE PORTFOLIO STANDARDS AND SOLAR RENEWABLE ENERGY CERTIFICATES

Energy Science, Engineering and Technology

Additional books in this series can be found on Nova's website under the Series tab.

Additional E-books in this series can be found on Nova's website under the E-book tab.

Energy Policies, Politics and Prices

Additional books in this series can be found on Nova's website under the Series tab.

Additional E-books in this series can be found on Nova's website under the E-book tab.

ENERGY SCIENCE, ENGINEERING AND TECHNOLOGY

SOLAR POWER, RENEWABLE PORTFOLIO STANDARDS AND SOLAR RENEWABLE ENERGY CERTIFICATES

VANESSA M. EVANS
AND
WADE L. REED
EDITORS

Nova Science Publishers, Inc.
New York

NOTICE TO THE READER

Library of Congress Cataloging-in-Publication Data

ISBN: 978-1-62081-532-8

Published by Nova Science Publishers, Inc. † New York

CONTENTS

PREFACE

Among the available options for encouraging the increased deployment of renewable electricity, renewables portfolio standards (RPS) have become increasingly popular. The RPS is a relatively new policy mechanism, however, and experience with its use is only beginning to emerge. One key concern that has been voiced is whether RPS policies will offer adequate support to a wide range of renewable energy technologies and applications or whether, alternatively, RPS programs will favor a small number of the currently least-cost forms of renewable energy. This book documents the design of and early experience with state-level RPS programs in the United States that have been specifically tailored to encourage a wider diversity of renewable energy technologies, and solar energy in particular. State-level RPS programs specifically designed to support solar energy have already proven to be an important, albeit somewhat modest, driver for solar energy deployment, and those impacts are projected to continue to build in coming years.

Chapter 1 - Among the available options for encouraging the increased deployment of renewable electricity, renewables portfolio standards (RPS) have become increasingly popular. The RPS is a relatively new policy mechanism, however, and experience with its use is only beginning to emerge. One key concern that has been voiced is whether RPS policies will offer adequate support to a wide range of renewable energy technologies and applications or whether, alternatively, RPS programs will favor a small number of the currently least-cost forms of renewable energy. This report documents the design of and early experience with state-level RPS programs in the United States that have been *specifically* tailored to encourage a wider diversity of renewable energy technologies, and solar energy in particular. As shown here, state-level RPS programs specifically designed to support solar have already proven to be an important, albeit somewhat modest, driver for solar energy deployment, and those impacts are projected to continue to build in the coming years. State experience in supporting solar energy with RPS programs is mixed, however, and full compliance with existing requirements has not been achieved. The comparative experiences described herein highlight the opportunities and challenges of applying an RPS to specifically support solar energy, as well as the importance of policy design details to ensuring that program goals are achieved.

Chapter 2 - Solar renewable energy certificates (SRECs) represent the generation attributes of solar energy systems and can be traded separately from commodity electricity. A SREC is created for each megawatt-hour of electricity generated from solar energy systems. SREC markets have emerged in nine U.S. states and Washington, D.C., as a method to meet compliance with solar carve out provisions of renewable energy portfolio standards (RPS). Solar carve out provisions set requirements for solar generation or capacity. This report examines the status and trends of SREC markets. Key issues and trends discussed in the report include:

- SREC markets are relatively young but expected to grow rapidly in coming years as state solar requirements ramp up. New Jersey has been the dominant market for SREC trading to date, but SREC markets in other states are expanding. Nationwide, solar carve outs in SREC markets are scheduled to grow from more than 520 MWAC in 2011 to nearly 7,300 MWAC in 2025.
- Policymakers establish key elements when designing SREC policies. Some states allow solar thermal to participate in SREC markets, but overall the SREC market is dominated by solar photovoltaics (PV). In terms of geographic eligibility, several states limit eligibility to in-state development, while a few allow SRECs from a broader geographic region.
- Solar alternative compliance payments (ACPs) generally set a ceiling on prices and are scheduled to decline over time, reflecting expectations of declining PV costs. In shortage situations, SRECs typically trade near the solar ACP.
- Lack of long-term contracts has been a barrier to project developers obtaining financing in some markets. This issue may continue to be a problem particularly in areas where rebates or other supplemental incentives expire and SRECs become the primary incentive stream for financing systems. Several states have instituted provisions such as

long-term contracting requirements, price floors, or standardized long-term contracts to try to overcome these challenges.

- Rate caps exist in some form in five SREC markets but have not been reached or are yet to be evaluated. The modest size of most solar carve outs suggests that it is not likely that rate caps will be reached in the near term.
- SREC markets are supporting a mix of PV system sizes, including residential, nonresidential, and utility-scale systems. In most markets, about two-thirds of registered projects are smaller than 10 kW. Projects larger than 250 kW dominate capacity installed in a few markets, including North Carolina, Massachusetts, and New Jersey. There has been a trend toward large-sized projects in recent years.
- Compliance has proved challenging in a number of SREC markets in recent years, but adequate supplies are expected going forward in many of these regions. A handful of states reported compliance shortfalls in 2009. The first compliance periods for a few markets occurred in 2010 or 2011. Recently, SREC spot prices have leveled or decreased in a number of markets as supplies have increased and compliance challenges have eased.
- SREC markets are expected to grow to more than 4,300 MW by 2020. In the near term, a sharp drop in SREC prices indicates that some markets may be oversupplied. An increase in installations has occurred in the last year due to a greater focus on U.S. markets, in large part due to slower growth in Europe, as well as a substantial drop in module prices. Capacity growth has been fueled in part by the approaching end of the Federal Treasury 1603 cash grant program, which will expire at the end of 2011. Looking forward, the largest markets to date are expected to slow over the next year or two but others are expected to continue to expand in the interim.

In: Solar Power, Renewable Portfolio Standards ... ISBN 978-1-62081-532-8
Editors: V. M. Evans and W. L. Reed

Chapter 1

SUPPORTING SOLAR POWER IN RENEWABLES PORTFOLIO STANDARDS: EXPERIENCE FROM THE UNITED STATES*

Ryan Wiser, Galen Barbose and Edward Holt

ABSTRACT

Among the available options for encouraging the increased deployment of renewable electricity, renewables portfolio standards (RPS) have become increasingly popular. The RPS is a relatively new policy mechanism, however, and experience with its use is only beginning to emerge. One key concern that has been voiced is whether RPS policies will offer adequate support to a wide range of renewable energy technologies and applications or whether, alternatively, RPS programs will favor a small number of the currently least-cost forms of renewable energy. This report documents the design of and early experience with state-level RPS programs in the United States that have been *specifically* tailored to encourage a wider diversity of renewable energy technologies, and solar energy in particular. As shown here, state-level RPS programs specifically designed to support solar have already proven to be an important, albeit somewhat modest, driver for solar

* This is an edited, reformatted and augmented version of U.S. Department of Energy, Office of Energy Efficiency and Renewable Energy, Office of Electricity Delivery and Energy Reliability and the National Renewable Energy Laboratory and the Clean Energy States Alliance, Publication No. LBNL- 3984E, dated October 2010.

energy deployment, and those impacts are projected to continue to build in the coming years. State experience in supporting solar energy with RPS programs is mixed, however, and full compliance with existing requirements has not been achieved. The comparative experiences described herein highlight the opportunities and challenges of applying an RPS to specifically support solar energy, as well as the importance of policy design details to ensuring that program goals are achieved.

EXECUTIVE SUMMARY

Among the available options for encouraging the increased deployment of renewable electricity, renewables portfolio standards (RPS) have become increasingly popular. By design, most RPS policies were originally developed to be largely technology-neutral, stimulating competition among various renewable energy technologies and allowing the most economically attractive technologies to win out. One concern that has been voiced, however, is whether RPS policies of this design will offer "adequate" support to a wide range of renewable energy technologies and applications or whether, alternatively, such RPS programs will favor a small number of the currently least-cost forms of renewable energy. This report documents the design of and early experience with state-level RPS programs in the United States that have been *specifically* tailored to encourage a wider diversity of renewable energy technologies, and solar energy in particular. Key findings from this analysis are as follows:

State RPS programs have not yielded a significant diversity of renewable resources thus far, though there are signs that this may be changing in some regions. As of the end of September 2010, 29 states and the District of Columbia had established binding RPS policies. Wind power has been the primary resource installed as a result of these policies, representing an estimated 94% of all RPS-driven renewable energy capacity additions in the U.S. from 1998-2009. Going forward, however, a more diverse set of renewable energy technologies may play a larger role in meeting RPS compliance obligations. In part, this is a consequence of the growing prevalence of RPS policy designs aimed specifically at encouraging greater resource diversity. At the same time, the trend towards greater renewable resource diversity is also being driven by the improved economics of solar relative to wind power, particularly in California and the desert southwest, where a sizeable pipeline of utility-scale solar projects has been announced. Though most of these projects are not yet in operation, these announcements, along with the growing use of RPS designs specifically tailored to increase resource diversity, portend a very real prospect for increased renewable resource diversity within state RPS programs, particularly in regions that contain high-quality solar resources and/or limitations in the supply of other forms of renewable energy.

Many states have adopted RPS policy designs explicitly aimed at supporting greater renewable resource diversity, and solar energy in particular. In addition to the suite of policies that might be used *in addition to* an RPS to support greater renewable resource diversity, at least two RPS design options exist that can be and are being used to encourage or even require greater resource diversity *within* RPS programs. First, RPS policies can incorporate **set-asides** consisting of different targets for different renewable technologies or applications, often with varying penalty levels to further encourage compliance. Alternatively, or in addition, preferred technologies or applications can be supported through **credit multipliers** of various designs that give preferred resources additional credit towards meeting a supplier's RPS compliance obligation. Of the 30 U.S. jurisdictions with RPS programs, the vast majority – 26 in total – have developed RPS programs with set-asides and/or credit multipliers, and in many cases these provisions have been adopted relatively recently. Most common are set-asides for solar energy (14 states) or distributed generation (4 states). However, set-asides for other renewable energy technologies (or broader technology bands) and for "community" renewable energy projects have also been established in a number of states. Credit multipliers of various magnitudes and durations are currently employed for solar energy (5 states), wind energy (3 states), and other sources. The greater popularity of set-asides suggests that state policymakers in the U.S. have tended to judge the advantages of set-asides to be greater than those of multipliers.

The design of solar and distributed generation (DG) set-asides varies widely across states. Design variations reflect the differing objectives of state policymakers, state-specific political exigencies, and electricity market designs and regulatory frameworks. As a result, a patchwork of solar support policies exists, demonstrating both a range of design choices available to policymakers, and how those design choices can impact policy outcomes. Key design issues highlighted and discussed in this report include: solar and DG targets and timeframes, eligibility rules and the use of multipliers to target particular preferred types of solar technologies or applications, contracting requirements and incentive mechanisms to overcome financing barriers, protocols for the measurement and tracking of solar RECs (SRECs), and cost containment and compliance enforcement.

Solar and DG set-asides have played a significant role in the recent growth of the U.S. solar market. Solar capacity in the U.S. has expanded rapidly over the past several years, albeit from a relatively small base. This growth has been driven by a number of factors, including – though not limited to – state RPS programs with solar or DG set-asides. As one relatively simple indicator of the impact of these polices, in each year from 2005-2009, 65-81% of the annual grid-connected PV capacity additions in the U.S. outside of California occurred in states with active or impending solar/DG set-aside obligations.[1] In aggregate, from 2000 through 2009, more than 250 $_{\mathrm{MWac}}$ of PV capacity is estimated to have been brought on-line to meet state-level solar or DG set-asides. The fact that solar capacity growth outside of California has centered, by and large, in states with solar or DG set-asides, suggests that these policies have played a key role in accelerating solar deployment in the U.S. These impacts have been most apparent in New Jersey (101 $_{\mathrm{MWac}}$ through 2009), Colorado (46 MW_{ac}), Arizona (36 MW_{ac}), New York (22 MW_{ac}), and Nevada (19 MW_{ac}); these states were also the largest PV markets in the U.S., after California, from 2007-2009. The impact of solar set-asides is also evident in the budding resurgence of the concentrating solar power (CSP) market in the U.S. After initial growth in the late 1980's, construction of CSP capacity in the U.S. largely ceased. This has begun to change in recent years, with two new CSP projects constructed to meet solar set-aside requirements: the 1 MW Saguaro project installed in Arizona in 2006 and the 64 MW Solar One facility installed in Nevada in 2007. Although experience with credit multipliers is limited, the available evidence suggests that, unlike set-asides, they have not yet resulted in a significant increase in solar generation.

Compliance with solar/DG set-aside targets has been mixed, highlighting the importance of careful policy design. Across the nine states with active solar/DG set-aside obligations in 2008 (the most recent year for which comprehensive set-aside compliance data are available), 68% of the aggregate solar/DG set-aside compliance obligation was achieved through the purchase of qualifying renewable energy and/or RECs. Though the targets in a number of these states amounted to only several megawatts of solar capacity, only three states (Colorado, Nevada, and Pennsylvania) fully achieved their solar/DG targets in 2008; the other six states (Arizona, Delaware, Maryland, New Jersey, New York, and Washington D.C.) met between 0% and 84% of their solar/DG set-aside obligations through the purchase of eligible forms of renewable energy and/or RECs. Though electricity suppliers in these states may not have been technically out of compliance with the solar/DG set-asides (due to alternative compliance payment mechanisms, funding limits, and force majeure provisions), these results demonstrate the challenges of meeting solar/DG set asides and the need for careful policy design.

The estimated retail rate impacts of solar/DG set-asides have thus far been relatively modest, though compliance costs have reached or are approaching 1% in some states. The retail rate impacts of solar/DG set-asides in 2009 have been estimated for seven U.S. states where data on average SREC prices or actual/budgeted expenditures are available.[2] For five of these states (Delaware, Maryland, New York, Ohio, and Pennsylvania), the estimated retail rate impacts are estimated at 0.01-0.04% of average retail electricity costs, reflecting the correspondingly low solar targets in those states, which ranged from 0.01-0.1% of retail sales in 2009. In New Jersey and Arizona, which had higher solar/DG set-aside targets of 0.2% and 0.3% of retail sales, respectively, estimated compliance costs reached 0.96% and 1.15% of total retail electricity costs.

State RPS programs, including both those with and without solar/DG set-asides, are poised to drive significant growth in the U.S. solar market. Solar capacity additions required to meet solar/DG set-aside targets will continue to grow as a greater number of existing set-asides take effect and as targets increase over time. By 2025, an estimated 9,500 $_{\text{MWac}}$ of solar generation capacity would be needed to fully meet the existing state set-aside requirements. Fully achieving set-aside targets during intervening years would require average annual capacity additions on the order of 400 $_{\text{MWac}}$ per year from 2010-2014 and approximately 600 $_{\text{MWac}}$ per year from 2015- 2025. In comparison, in 2009, approximately 107 $_{\text{MWac}}$ of solar capacity was added in order to meet solar and DG set-asides, a figure that would have been higher had full compliance with existing targets been achieved. At the same time, recent announcements for utility-scale solar projects suggest that future growth in solar capacity will, by no means, be limited to meeting solar or DG set-aside obligations. Much of this announced capacity, which totals more than 22,000 $_{\text{MWac,}}$ is planned for California or parts of the desert southwest and, if constructed, would largely serve to meet general RPS compliance obligations in the region. This latter trend suggests that, in the future, solar/DG set-asides may be especially important in states with poorer solar resources or for distributed and customer-sited solar applications.

A variety of emerging issues will affect the impact of RPS policies on solar growth. Meeting the long-term solar/DG set-aside targets, as identified in the preceding paragraph, will require a scale-up in the solar delivery infrastructure in many states. In addition to this challenge, a number of policy design issues may constrain the market's growth. In particular, many states have developed cost containment mechanisms that may ultimately become binding. Arizona and New York, for example, both missed their set-aside targets in 2008 as a result of funding limits, and Ohio utilities were granted a force majeure exemption in 2009, thereby curtailing achievement of that state's solar targets. Furthermore, some states, especially those with competitive retail electricity markets, continue to struggle with how to encourage long-term contracting for solar generation. In response, utilities and regulators in a number of states have sought to reduce SREC price risk and facilitate project financing through a variety of policy designs and other strategies, including: adopting long-term contracting requirements, offering financial incentives or financing programs, conducting centralized procurement of RECs, and developing other novel procurement models, including direct utility ownership of distributed PV assets. Further experimentation with such mechanisms may be needed to help ensure that solar set-aside targets are fully achieved in a least-cost fashion, especially in states open to retail electricity competition.

1. Introduction

Renewables portfolio standards (RPS) have, within the last decade, emerged as among the most popular forms of policy for supporting the deployment of renewable energy technologies. Though its design can and does vary, at its heart, an RPS simply requires that electricity suppliers (or, alternatively, electricity generators or consumers) purchase a growing quantity of renewable energy over time; most jurisdictions allow trade in renewable energy certificates to increase compliance flexibility and facilitate compliance verification. RPS programs – also known as quota systems, renewables obligations, renewable energy standards, or tradable green certificate programs – are widely used at the state level in the United States, and have been implemented in Australia, Belgium, Canada, China, India, Italy, Japan, Poland, Sweden, and the United Kingdom. Notwithstanding its breadth of adoption, the RPS is still a relatively new policy with limited experience. Despite its popularity, the RPS policy has its detractors. Though successfully employed in a number of instances, in some cases, RPS programs have done little to support growth in renewable energy supplies, and in others that growth has come at a high cost (see, e.g., IEA 2008; van der Linden et al. 2005). More generally, concerns have been expressed about the ability of RPS programs to support the long-term contracts that renewable energy generators require, the willingness of policymakers to enforce renewable energy purchase obligations, the complexity of crafting a well-designed program, the challenges of minimizing windfall profits for some generators, the need to enact supporting policies beyond the RPS, and whether alternative policies may deliver comparable social benefits at lower cost (see, e.g., Agnolucci 2007; Bergek and Jacobsson 2010; Butler and Neuhoff 2008; Chupka 2003; Cory and Swezey 2007; Finon and Perez 2007; Hvelplund 2001; Kildegaard 2008; Lauber 2004; Lipp 2007; Menanteau et al. 2003; Meyer 2003; Midttum and Gautesen 2007; Michaels 2008; Mitchell et al. 2006; Palmer and Burtraw 2005; Ringel 2006; Wiser et al. 2005a; Wiser et al. 2005b). By design, most RPS policies were originally developed to be largely technology-neutral, stimulating competition among various renewable energy technologies and allowing the most economically attractive technologies to win out. Consequently, a key concern sometimes voiced in academic and policy circles has been that an RPS of this design may not support the diversity of renewable energy technologies and applications that may ultimately be necessary in the long term if even-higher levels of renewable energy deployment are to be attained. Meyer and Koefoed (2003), Lauber (2004), and Ringel (2006), among others, discuss this concern in general, while Voogt and Uyterlinde (2006) show empirically that an EU-wide renewable energy target with uniform renewable energy certificate pricing is unlikely to provide much support for solar photovoltaic technologies in particular; similarly, analyses of Federal RPS proposals in the United States have typically found that wind and biomass are likely to dominate compliance, with lesser quantities of solar and geothermal energy (e.g., Kydes 2007; Nogee et al. 2007). Bergek and Jacobsson (2010) note the concern when evaluating Sweden's RPS, while Falconnett and Nagasaka (2010) conclude, as a result, that feed-in tariffs are superior in supporting less-mature technologies. van der Linden et al. (2005), meanwhile, note that small renewable energy producers may face barriers under an RPS due to the significant transaction costs and risks involved in participating in the resulting renewable energy market. What is sometimes missing from this literature, however, is a clear recognition that RPS programs can be specifically designed to encourage renewable resource diversity. At least two specific policy design approaches are feasible: developing technology- or application-specific set-asides (also called "bands" or "carve-outs"), or providing credit multipliers that offer preferred technologies or applications additional compliance credit towards the RPS.[3] A growing number of U.S. states have designed their RPS policies to apply one or both of these mechanisms to support a broader set of renewable energy technologies and applications; other countries have also, at times, employed similar mechanisms. Alternatively, RPS policies can be paired with other types of policies that provide targeted support to solar or other renewable technologies that may not fare well under traditional RPS designs. This report summarizes the design, early experience, and lessons learned from U.S. state RPS policies that have been specifically targeted at encouraging renewable resource diversity.[4] Because such policies in the U.S. have tended to emphasize support for solar energy, we also follow this emphasis. RPS programs are not the only, or perhaps even the most significant, form of support for solar energy either in the U.S. or internationally; feed-in tariffs and financial incentives also play major roles in solar deployment, and it is not our purpose to make the case for one policy over another. Instead, we focus exclusively on RPS policies because the growing popularity of such programs calls for a careful tracking and evaluation of policy design and results. Our focus on U.S. experience is driven by the fact that the U.S. arguably has the longest running and most diverse set of experiences with a wide range of RPS designs. The remainder of this report is structured as follows. We begin in the next section by providing background on state RPS programs, and on the early results of those programs in encouraging renewable energy generally, and renewable resource diversity specifically. We then turn to a general discussion of the RPS design approaches that can be used to encourage greater resource diversity broadly, and solar energy in particular. Following that, we (a) discuss in some detail the multiple ways in which solar-specific RPS provisions have been designed in the U.S., (b) summarize the early impacts of those policies on solar energy deployment and retail electricity rates, (c) document the level of compliance achieved to date with those policies, and (d) provide a forecast for how solar deployment may increase in the future under these policies if full compliance is achieved. We conclude the report by discussing major policy design considerations and summarizing our key findings.

2. State RPS Policy Background

2.1. Prevalence and Impact of State RPS Policies

Renewables portfolio standards have proliferated at the state level in the U.S. since the late 1990s. As of the end of September 2010, 29 states and the

District of Columbia had established binding RPS targets which, when fully implemented, will cover 56% of total U.S. retail electricity sales, requiring that a certain percentage of those sales be met with renewable energy; an additional seven states have passed legislation establishing voluntary renewable energy goals.[5] Although the design and final compliance targets of these policies vary widely, Figure 1 shows that many of the RPS programs require that eligible forms of renewable energy contribute 15-25% of retail sales by 2030 or sooner.

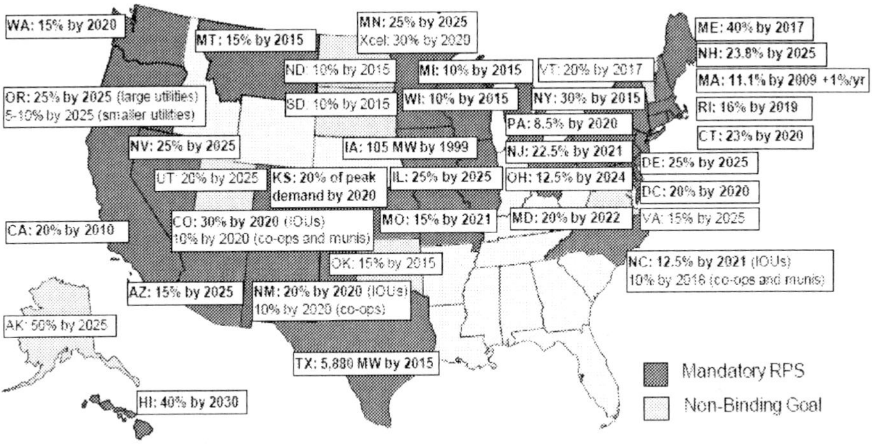

Source: Berkeley Lab.

Figure 1. Existing State RPS Policies and Non-Binding Renewable Energy Goals.

State RPS requirements have emerged as one of the most important drivers of renewable energy additions in the U.S., especially when coupled with other complementary policies. Of the more than 37 GW of non-hydro renewable energy capacity added from 1998 through 2009 in the U.S., roughly 61% (23 GW) occurred in states with an active or impending RPS compliance obligation.[6] In total, existing state RPS policies will require roughly 73 GW of new renewable capacity by 2025, representing roughly 6% of total U.S. retail electricity sales in that year and 30% of projected load growth between 2000 and 2025. If these states increase their renewable energy targets, as many have done in the past, or if additional states were to adopt RPS policies, the collective set of state RPS policies would require an even greater quantity of new renewable energy additions.

2.2. Impact of State RPS Policies on Resource Diversity to Date

Wind power has been the primary resource installed as a result of state RPS policies thus far, representing an estimated 94% of all RPS-driven renewable energy capacity additions in the U.S. from 1998-2009; the remaining 6% consists of biomass, solar, and geothermal (Figure 2). Though some regions of the country with RPS policies have experienced somewhat greater diversity in new renewable energy additions, wind has thus far consistently been the dominant renewable technology deployed as a result of state RPS programs (Figure 3). This is because, as a 'market-driven' mechanism, a traditional RPS tends to stimulate investments in lower-cost and lower-risk technologies; higher cost technologies will generally not be chosen during the competitive process.

Solar has historically tended to be less economically attractive than wind energy in most regions of the U.S., and therefore (until recently) has not generally been selected through RPS procurement processes. In addition, some renewable energy technologies and applications – solar PV included – have historically been primarily deployed as smaller-sized systems (in the case of PV, located on the customer-side of the meter), and have therefore faced a variety of other solicitation-related barriers to participating in RPS markets. In particular, smaller solar projects are not always able to easily participate in the solicitations that have become common under many state RPS programs given the high transaction costs of such participation, explicit minimum project size thresholds that sometimes exist, and/or stringent metering requirements that can apply. Finally, solar projects have also sometimes faced policy-related barriers to participation in RPS programs, for example, uncertainty over renewable energy certificate (REC) ownership for PV projects that receive cash incentives or net metering.

Going forward, a more diverse set of renewable energy technologies may play a larger role in meeting RPS compliance obligations, even in cases where RPS policies are not explicitly designed to encourage that diversity. In California, for example, between 2002 and September 2010, the state's investor owned utilities (IOUs) and publicly owned utilities (POUs) collectively signed contracts for 21 GW of new renewable capacity. Roughly 41% of this capacity consists of large new solar photovoltaic (PV) and concentrating solar power (CSP) facilities, with geothermal and biomass each representing another 3%; 53% is wind (CPUC 2010, CEC 2010). In addition, announcements of prospective large, utility-scale solar plants have been made under the RPS policies in Arizona, New Mexico, and Colorado. This shift towards utility-scale solar energy is largely being driven by the improved economics of solar relative to wind power, particularly in the desert southwest. Though most of these projects are not yet in operation, the announcements signal a very real prospect for increased renewable resource diversity within state RPS programs in areas of the country that contain high-quality solar resources and/or limitations in the supply of other forms of renewable energy.

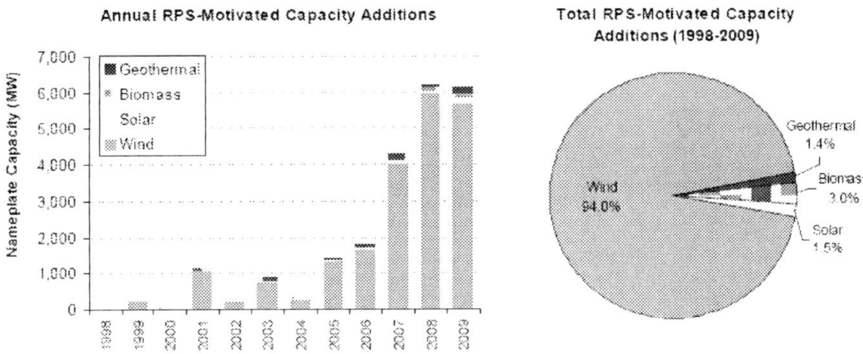

Notes: Renewable additions are counted as "RPS-motivated" if and only if they are located in a state with an RPS policy and commercial operation began no more than one year before the first calendar year with RPS compliance obligations. This approach ignores a number of complexities. First, most states allow out-of-state generation to count toward their RPS; thus, some renewable additions that, in fact, were motivated by RPS policies are excluded from the figure. Conversely, in some RPS states, renewable energy capacity has been added in recent years above and beyond the level necessary to meet those states' RPS targets, or has been added to serve voluntary green power markets; thus, some renewable additions included in the figure are not RPS-motivated.

Figure 2. Renewable Energy Additions in RPS States by Technology, 1998-2009.

2.3. RPS Policy Designs for Supporting Renewable Resource Diversity

Notwithstanding the future prospect for increased renewable resource diversity resulting from the narrowing cost differences among various renewable energy technologies, the fact remains that – to date – U.S. state RPS programs have tended to largely support utility-scale wind power projects. Traditionally designed RPS policies have not yet yielded significant solar additions, with the exception of larger, utility-scale solar projects and project proposals in the sunny desert southwest. Smaller, customer-sited solar applications have – thus far – fared particularly poorly when required to compete with larger, utility-scale wind power projects.

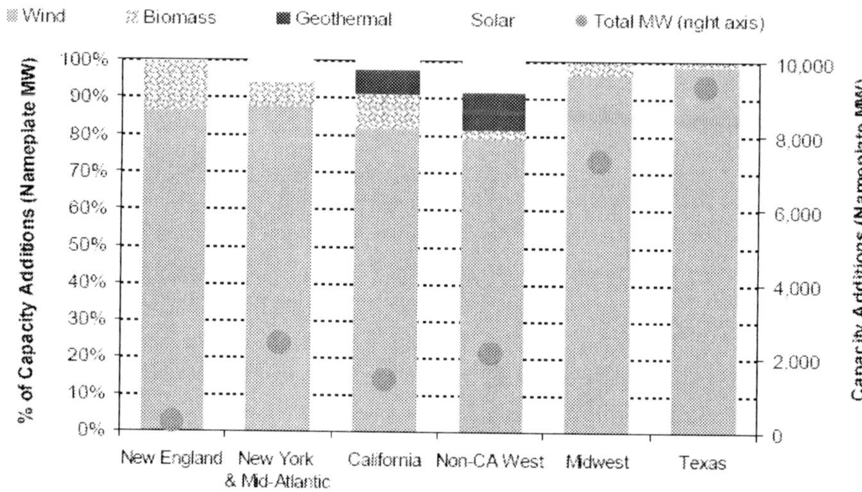

Notes: See Figure 2 notes for explanation of the methodology used to construct the figure.

Figure 3. RPS-Motivated Renewable Energy Additions by Technology and Region, 1998-2009.

In addition to the suite of policies that might be used *in addition to* an RPS to support greater resource diversity, however, at least two RPS design options exist that can be and are being used to encourage or even require greater resource diversity *within* RPS programs. First, RPS policies can incorporate **set-asides** (also called bands, tiers, or carve-outs) consisting of different targets for different renewable technologies or applications, often with varying penalty levels to further encourage compliance. Alternatively or in addition, preferred technologies or applications can be supported through **credit multipliers** of various designs that give preferred resources or applications additional credit towards meeting a supplier's RPS compliance obligation; in particular, generation from the designated technologies or applications, although issued one REC for each MWh, may be credited as more than one REC (depending on the multiplier) for RPS compliance purposes. In the former case, increased diversity is required through segmenting the overall RPS requirement into various sub-requirements, while in the latter case increased diversity is encouraged but not strictly obligatory.

Set-asides and credit multipliers each offer their respective sets of corresponding advantages and disadvantages, whether applied to solar or to other preferred technologies or applications. Focusing on solar (see Table 1), chief among the advantages of set-asides is that they provide a degree of certainty that the RPS will result in the development of a specific amount of solar resources: a key ingredient to financing large-scale solar projects. Credit multipliers, in contrast, provide no such certainty, and to the extent that they do stimulate solar development, they do so at the expense of reducing the effective RPS percentage.[7] In addition, set-asides can provide policymakers with a greater ability to create a tailored form of support for solar to address solicitation-related barriers – for example, by establishing solar-specific procurement mechanisms and REC tracking protocols.

Table 1. Advantages and Disadvantages of Solar Set-Asides and Solar Credit Multipliers

Set-Asides	Credit Multipliers
Advantages	Advantages
• Recognizes unique benefits of solar • Greater certainty in the amount of solar deployment • Targets cost and solicitation barriers • Does not reduce effective RPS percentage • Less risk of over-subsidization of solar • May enable larger market for solar than multiplier depending on details	• Recognizes unique benefits of solar • Lower risk of higher/uncertain cost impact • Allows policymakers to clearly signal the degree to which solar is valued relative to other resources • Does not "pick winners" as directly as set-asides • May enable larger market for solar than set-aside depending on details
Disadvantages	Disadvantages
• Greater risk of higher/uncertain cost impact • May cause overall RPS cost cap to be reached, if a separate cost cap for set-aside is not established • Picks winners more directly than multiplier • Establishing level of set-aside is challenging • Set-aside often rigidly set without easy ability to alter given changes in market conditions	• Does not ensure certain amount of solar deployment • Does not directly target solicitation barriers for smaller solar projects • Reduces effective RPS percentage • Establishing multiplier value at "correct" level over time is challenging: requires supervision • If multiplier is not reduced as costs decline, could lead to over-subsidization

A prime disadvantage of set-asides, on the other hand, is the risk that they will put upward pressure on RPS compliance costs, if solar is more expensive than other sources of renewable generation. Credit multipliers do not entail this risk, as solar resources would be used to meet the RPS target only to the extent that doing so is lower cost than the alternate renewable options available. Credit multipliers also have the advantage of providing a means by which states can directly indicate the degree to which they value particular types of renewable resources over others. It is perhaps in recognition of the unique advantages offered by both set-asides and multipliers that several states have adopted policies combining both mechanisms, though the greater popularity of set-asides suggests that state policymakers in the U.S. have tended to judge the advantages of set-asides to be greater than those of multipliers.

Table 2. Application of Set-Asides and Multipliers in State RPS Programs

	Set-Asides		Credit Multipliers
General Technology	Specific Technology	Specific Application	
Class I vs. II:CT, DC, MA, MD, ME, NJ	•Solar Energy*: DC, DE, IL, MA, MD, MO, NC, NH, NJ, NM, NV, OH, OR, PA •Wind Energy: IL, ME (goal), MN, NJ (offshore), NM •Existing Biomass/Methane: NH •Existing Hydropower: NH •Geothermal or Biomass: NM •Swine Waste: NC •Poultry Waste: NC •Non-Wind: TX (goal)	•Distributed Generation: AZ, CO, NM, NY •Community Ownership: MN (goal), MT (wind), OR (goal, community and small scale)	•Solar Energy: CO (POUs), DE (general RPS), MI, NV (PV), OR •Wind Energy: DC, MD, DE (offshore) •Methane: DC, MD •Distributed Generation:NV (PV), WA •Community Ownership: CO, ME •Fuel Cells: DE •Waste Tires: NV •Non-Wind: TX

* Rhode Island passed legislation in 2009 requiring the state's electric distribution companies to sign long-term renewable energy contracts totaling 90 MW by 2014, of which 3 MW must be solar energy. The long-term contracting requirement is not technically part of the state's RPS, and we therefore do not include Rhode Island in our tally of states with RPS solar set-asides.

Of the 30 U.S. jurisdictions with RPS programs, the vast majority – 26 in total – have designed their RPS programs with technology and/or application diversity in mind, and both set-asides and credit multipliers have been common.[8] As shown in Table 2, these policies take many forms, suggesting that policy interests and preferred resources are far from uniform across states. Most common are set-asides for solar energy (14 states) or distributed forms of renewable generation more broadly (4 states). However, a number of states have adopted set-asides for wind and other renewable energy technologies, or for "community" renewable project ownership. Other states have developed broader technology bands that separate renewable energy sources into "Class I" and "Class II" technologies, where Class I technologies typically represent the newer, more costly, and/or more-preferred sources. Finally, a number of states have developed credit multipliers of various magnitudes and durations for solar energy (5 states), wind energy (3 states), and other resources.

3. SUPPORTING SOLAR IN AN RPS: SET-ASIDE DESIGN VARIATIONS

3.1. Prevalence of Solar-Specific RPS Designs

Though the approaches used are varied, solar energy has been the most common target of RPS policy designs aimed at promoting renewable resource diversity. As of the end of September 2010, 14 of the 30 RPS policies contained solar-specific set asides, while four states had developed distributed generation (DG) set-asides that will likely serve, in large measure, to support solar (Figure 4). In addition, five states had adopted solar multipliers, either in lieu of or in combination with set-asides, and two additional states had multipliers for DG.

Many of these policies have been enacted recently, demonstrating a growing concern among many states about the tendency of traditionally designed RPS programs to provide little support for certain solar applications; 10 of the 17 jurisdictions with solar or DG set asides, for example, have developed those policies just since the beginning of 2007. Also of note is a general trend away from multipliers over time; based on limited historical experience, multipliers have not proven particularly effective at stimulating solar adoption, and states have increasingly sought the greater certainty of solar deployment associated with obligatory set-asides. For this reason, the remainder of this section exclusively covers those states with solar or DG set-aside policies.

As implied by Figure 4, the design of state solar and DG set-asides varies substantially; uniformity in design has neither been sought, nor achieved. Design variations reflect the differing objectives of state policymakers, state-specific political exigencies, and electricity market designs and regulatory frameworks. As a result, a patchwork of solar support policies exists, demonstrating both a range of design choices available to policymakers, and how those design choices can impact policy outcomes. Key design issues are discussed further below and include: solar targets and timeframes, eligibility rules and the use of multipliers to target particular preferred types of solar technologies or applications, contracting requirements and incentive mechanisms to overcome financing barriers, protocols for the measurement and tracking of solar RECs (SRECs), and cost containment and compliance enforcement.

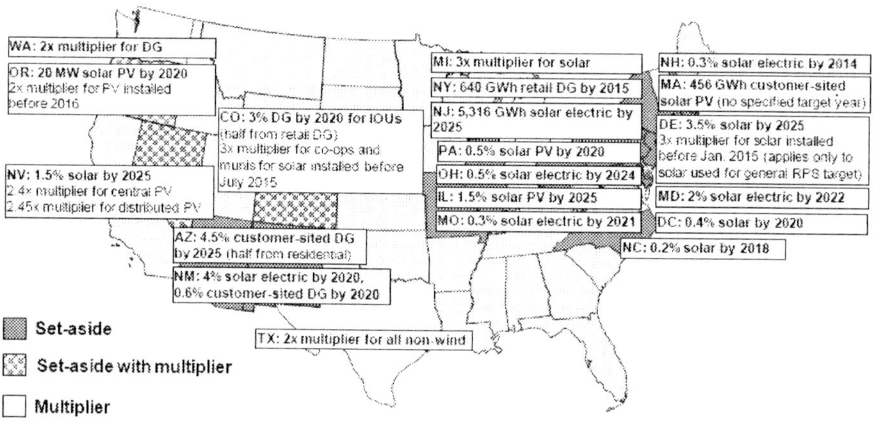

Figure 4. Set-Asides and Multipliers for Solar Energy and Distributed Generation in State RPS Policies.

3.2. Solar Targets and Timeframes

Among the multitude of design variations, perhaps the most significant is that the ultimate percentage targets under solar and DG set-asides varies widely, with some states eventually requiring solar or DG generation of 1-6% of the retail electricity sales on which the RPS applies, while others have targets that rise to only a fraction of a percent.[9] The ultimate targets are listed in Figure 4, while Figure 5 presents data on how the targets increase over time; the latter shows that growth trajectories vary, as does the frequency of upward ratchets over time.[10] Although a number of states with relatively aggressive percentage targets are located in the Southwestern U.S. (Arizona, New Mexico, Colorado, and Nevada), with its robust solar resource, this is by no means exclusively the case. In fact, the highest percentage solar target (6%) belongs to New Jersey, and a number of other Mid-Atlantic or Mid-Western states (Delaware, Maryland, and Illinois) have adopted solar targets that ultimately reach 1.5-3.5% of retail electricity sales by 2025 or sooner.

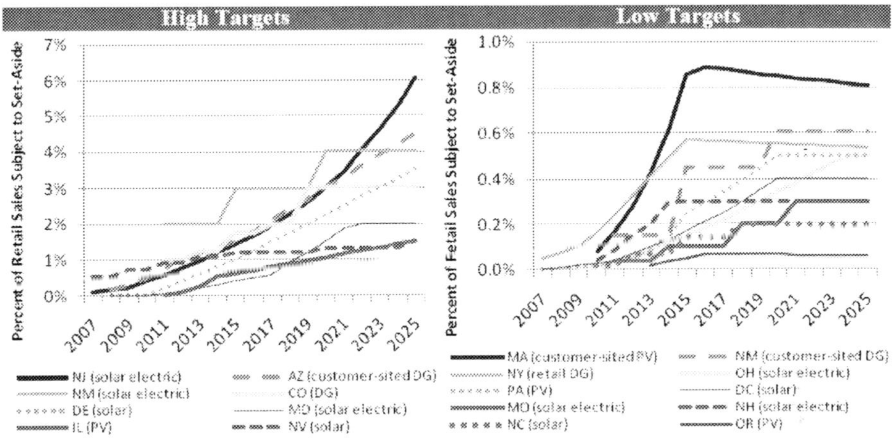

Figure 5. State RPS Set-Aside Targets for Solar and Distributed Generation.

3.3. Eligibility Rules and Multipliers

The aggressiveness of the set-aside obligations must also be viewed within the context of differences in resource eligibility. Here, there are many design options to consider: whether to focus on solar or to be more inclusive; what types of solar technologies and applications are eligible; whether to limit eligibility to customer-sited solar or to include utility-scale solar; and whether to place geographic limitations on solar eligibility. As with most elements of program design, the answers depend on a clear understanding of policy goals.

As shown in Table 3, states have taken very different approaches to defining eligible technologies and applications for their set-asides. Specifically, solar electric technologies (just PV, or PV and CSP) are generally the focus of solar set-asides, though solar heating and/or cooling (SHC) technologies, which may include solar hot water, are also eligible for solar/DG set-asides in six states.[11] Four states have broader distributed generation set-asides in which solar is one of several eligible technologies, while other states have solar set-asides that specify that a certain proportion of the target must come from specific applications, e.g., customer-sited distributed solar or residential solar.

A number of the policies require that projects eligible under the relevant set-aside be located instate, especially customer-sited DG. In some states, this requirement is effectuated by requiring that eligible projects be connected to the distribution system of an electric utility serving customers in the state.

Table 3. Technology and Application Eligibility for Solar and DG Set-Asides

State	PV	CSP	SHC	Non-Solar DG	Eligible Applications	Geographic Restrictions
AZ	✔	✔	✔	✔	Customer-sited; 90% retail DG (installed behind customer-meter); ½ residential; SHC must displace electricity	Effectively in-state(a)
CO	✔	✔		✔	≤ 30 MWdc per project, ½ retail DG	In-state required for retail DG; in-state multiplier for all other RPS resources, including wholesale DG
DC	✔	✔	✔		All	Out-of-District solar accepted only if insufficient in-District(b)
DE	✔	✔	✔		SHC must displace electricity	Delivery required to region
IL	✔				All	Preference for in-state (IOUs only)
MA	✔				Customer-sited and ≤ 6 MWdc	In-state
MD	✔	✔			All	Out-of-state accepted until 2012 only if insufficient in-state
MO	✔	✔			All	In-state multiplier for all RPS resources, including solar; deliveryrequirements under development
NC	✔	✔	✔		All	Unbundled RECs allowed with limits; otherwise delivered to state
NH	✔	✔			All	Delivery required to region
NJ	✔	✔			All	Effectively in-state(c)
NM	✔	✔		✔	DG set-aside: Customer-sited Solar set-aside: All	DG set-aside: Effectively in-state(d) Solar set-aside: Delivery required to state and preference for resources located in-state
NV	✔	✔	✔		All	Delivery required to state
NY	✔		✔	✔	Retail DG; SHC must displace electricity	In-state

State	PV	CSP	SHC	Non-Solar DG	Eligible Applications	Geographic Restrictions
OH	✔	✔			All	Delivery required to state; ½ of all RPS resources must be in-state
OR	✔				500 kW to 5 MW per project	In-state
PA	✔				All	In region

(a) Arizona's RPS rules require that distributed solar energy resources are sited at a customer facility and serve on-site customer load or provide wholesale capacity and energy to the local utility distribution company.

(b) Washington D.C.'s RPS rules require that solar resources are connected to the distribution system serving the District, unless insufficient in-District solar is available.

(c) New Jersey's RPS rules require that solar resources are connected to the in-state distribution system.

(d) New Mexico's RPS rules require that DG resources serve on-site customer load or serve customers in contiguous distribution substation areas.

Other states allow out-of-state solar facilities to qualify, but that allowance sometimes comes with limitations aimed at encouraging in-state development. In Maryland, for example, retail electricity suppliers are allowed to purchase SRECs from out-of-state solar projects only until 2012, and only if sufficient in-state SRECs are not available. At the same time, customer-sited solar facilities located in Maryland that wish to sell their SRECs are first obliged to offer those SRECs for sale to electricity suppliers with compliance obligations under the state's RPS. Ultimately, decisions on the treatment of out-of-state facilities requires a balance between the desire of policymakers to promote in-state economic development, on the one hand, and the need to not violate the interstate commerce clause of the U.S. Constitution prohibiting states from erecting barriers to interstate trade, on the other.[12]

Finally, some states have established credit multipliers, in concert with set-asides, to try to direct investments towards certain preferred solar technologies, applications, or project locations. Oregon currently employs a time-limited multiplier for solar to motivate accelerated deployment. Nevada, on the other hand, provides multipliers that seek to motivate PV applications over CSP applications, whereas Colorado and Missouri offer multipliers for in-state facilities (applicable only to the wholesale DG portion of the Colorado's DG set-aside).[13] Several other states currently employ multipliers for solar energy (or that apply to solar energy) without an associated set-aside (Michigan, Texas, Washington, Colorado[14]); similarly, Delaware offers a time-limited multiplier for in-state solar (and fuel cells) that applies to solar resources that used to meet the general RPS target, but not the solar set-aside.

3.4. Policy Design Options for Facilitating Project Financing

RPS policies are, in theory, intended to motivate the private sector to meet renewable energy purchase obligations at least cost and with a minimum of ongoing regulatory oversight. In practice, however, state policymakers have often guided the process of resource selection and procurement due to concerns that electricity suppliers either may not pursue least-cost compliance paths, or may otherwise not seek to procure certain preferred solar resources or applications. In the former case, the concern is typically that electricity suppliers may be unwilling or unable to enter into the longer-term contracts that renewable generators require for financing and that may lead to lower overall compliance costs than shorter-term purchases. This is especially true in restructured electricity markets open to retail competition where retail suppliers do not have certainty over their future load obligations, and therefore also lack certainty on the size of their future renewable energy purchase mandate. In the latter instance, the concern is often one of seeking to support a wide range of solar applications, including residential and commercial solar installations that might not otherwise compete with central station solar projects, or that may have a tendency to need up-front payment mechanisms.

Minimum contract duration requirements are one approach that states have taken to improving certainty in REC revenue for renewable energy projects and facilitating project financing. Among states with solar or DG set-asides, Colorado, Maryland, and Nevada have established minimum contract duration requirements for all solar energy contracts of 20, 15, and 10 years, respectively. In addition, although North Carolina does not have a fixed contract duration requirement, the state does require that contracts with solar energy facilities "be of sufficient length to stimulate development of solar energy." In practice, contracting requirements are often simpler to implement in markets still characterized by regulated, vertically integrated utilities because (a) regulators have a history of strong oversight of energy procurement decisions in these contexts; (b) these utilities can be relatively certain about their future retail loads and RPS obligations because they are not subject to competition from other suppliers; and (c) with traditional cost-based regulation, cost-recovery can be assured.

Contracting requirements are more difficult to apply in restructured electricity markets in which retail competition is allowed, because electricity suppliers face greater uncertainty about their future load and RPS obligations, and because the state may lack the legal authority or desire to impose contracting requirements on competitive retail suppliers. As such, only one restructured state (Maryland) requires long-term contracting for SRECs by competitive suppliers, and this requirement is imposed only if the retail electricity supplier purchases SRECs directly from the solar generator (not through a broker or other intermediary). Other restructured states have, instead, turned to a variety of alternate models to facilitate revenue certainty for solar (and other renewable) projects required under their RPS programs. New York and Illinois have both adopted a central procurement model, whereby a state agency issues solicitations for RECs on behalf of obligated electricity suppliers (see Text Box 1). In Delaware, legislation was passed that suspends the three-year lifetime of RECs that are held by state's Sustainable Energy Utility (SEU), which serves as a REC aggregator; this provision was intended to allow the SEU to hold RECs off the market during periods of oversupply and thereby support greater price stability. New Jersey and Massachusetts have each developed alternative approaches aimed at supporting the financing of solar energy projects specifically. In New Jersey, the state's regulated electric distribution utilities effectively act as SREC wholesalers by purchasing SRECs under long-term contract and then re-selling them to competitive retail electricity suppliers; in Massachusetts, the state plans to conduct annual SREC auctions to provide a backstop SREC market for projects that do not arrange bilateral SREC contracts (see Text Box 2 for further details).

In addition to or instead of the approaches described above, many states or utilities have developed standard offer SREC purchase programs and/or incentive programs (under which SRECs may or may not be transferred to the utility or state). Like long-term contracting requirements, incentive and standard offer programs of this type – whether the payment is made up-front in the form of a rebate or over time based on project performance – mitigate revenue uncertainty for the solar project owner, but such programs can also advance other objectives, such as providing differentially higher support for certain types of solar applications (e.g., customer-sited or residential systems). In a number of jurisdictions (Colorado, Missouri, Nevada, New York, and the District of Columbia), these contracting and incentive programs were created specifically as a mechanism for meeting solar or DG set-aside requirements.[15] However, standard offer SREC purchase programs or incentive programs have been developed voluntarily in most other states, independent of any explicit requirement included with the state's RPS legislation. In general, these programs target customer-sited PV projects, and in some instances include SHC technologies as well. Often, eligible applications are limited to projects less than a designated size (or are limited to net-metered systems, where net metering rules place restrictions on system size) and/or the total per-system payment amount is capped. Depending on the program and project size, payment may be provided up-front or over time, based on measured solar energy production. If SRECs are transferred from the system owner to the utility or state in exchange for payment, the SREC transfer may extend for a designated period of time (e.g., 3 years in the case of New York's PV incentive program or 20 years in the case of Xcel Energy's Solar Rewards Program), or it may continue indefinitely (e.g., in Arizona Public Service Company's Renewable Energy Incentive Program and Nevada Energy's SolarGenerations Program).

Finally, utilities and state regulators have also addressed the challenges associated with solar project financing though direct utility ownership of solar assets – focusing particularly on distributed PV applications (see Text Box 3 for examples). This trend – which exists both in states with solar or DG set-asides and in states with no such policies – reflects a variety of objectives, in addition to facilitating project financing. At the broadest level, the challenges of meeting aggressive RPS targets and the desire to diversify RPS compliance risk have prompted utilities to pursue a multitude of procurement strategies. Solar asset ownership gives utilities greater control over meeting their RPS targets, and procurement models focusing on distributed solar, specifically, allow utilities to side-step the transmission- and permitting-related barriers that can slow the development of central-station generation. Moreover, direct ownership of solar assets offers utilities the opportunity to rate-base capital costs and thereby earn a return on solar asset ownership – an opportunity made more financially attractive by changes to federal tax law in 2008 allowing utilities to claim the Section 48 business solar investment tax credit. From the customer perspective, hosting a PV facility that is owned by the utility obviates the need for the customer to raise the funds needed to cover the high capital cost of a solar facility, and lowers the customers' transaction costs of evaluating, selecting, and dealing with private contractors. A disadvantage of greater direct utility involvement in the solar market may be the natural tendency of utilities, based on economies of scale, to focus on larger commercial and industrial solar applications to the detriment of residential and small commercial customers. There may also be concerns about utilities competing with private sector solar services, although utilities can still work with solar industry trade allies to identify, develop, install, and service the installations under contractual arrangements. Allowing or requiring that the utility acquire a portion of the solar capacity through competitive procurement from independent power producers is one option to address this latter concern.

Text Box 1: Central Procurement of RPS Resources

One way to address issues caused by uncertain RPS obligations for retail electricity suppliers in restructured markets is to appoint a central procurement agent, thereby relieving the competitive suppliers of direct compliance obligations. Two states – New York and Illinois – have thus far adopted a central RPS procurement model, applicable both to overall RPS compliance obligations and to the solar or DG set-aside obligations.

New York: Investor-owned distribution utilities in New York collect a surcharge on retail sales for RPS implementation. The New York State Energy Research and Development Authority (NYSERDA) manages this RPS fund and uses it both to purchase RECs from new qualifying renewable energy generators through a competitive process and also to provide financial incentives for customer-sited distributed renewable resources, including solar. Under the rules for receiving the latter incentives, customers with DG systems agree to relinquish the first three years of RECs associated with their systems to NYSERDA, and the aggregate generation from such DG facilities is counted towards the state's DG RPS set-aside.

Illinois: The Illinois Power Agency (IPA) was created to develop electricity procurement plans for the state's large IOUs. For the purposes of RPS compliance, the IPA plans and administers elements of the competitive procurement processes for renewable energy; the IPA also prepares a compliance plan for meeting RPS requirements. Though the IPA manages the competitive solicitations for renewable energy on behalf of the state's large IOUs, unlike in New York, it is up to the individual utilities to contract with the winning bidders of the IPA-managed solicitations. Until recently, all of the IPA RPS procurements have been for unbundled RECs, and just for one-year periods. Starting in 2010, however, the IPA will conduct a 20-year procurement for RECs bundled with an energy swap. The RECs under the 2010 solicitation are required to come from specified facilities, with energy delivered to a utility delivery point.

As yet, procurement plans have not differentiated solar from other RPS resources, because the solar requirement does not begin for the state's IOUs until 2011-2012. The IPA also manages a portion of the RPS procurement for the state's competitive retail suppliers, which are required to meet at least 50% of their annual RPS procurement obligation by providing funds (termed Alternative Compliance Payments, though similar to a systems benefit charge) to the IPA, which uses those funds to purchase RECs. Solar set-aside obligations as applied to the competitive suppliers do not begin until 2015, however, so as with the IOUs, the IPA has not yet issued a solar-specific solicitation.

Text Box 2. Supporting Solar Project Financing in New Jersey and Massachusetts

New Jersey and Massachusetts have each developed innovative, but distinctly different, approaches to addressing the financing challenges for solar energy projects in restructured electricity markets.

New Jersey: As part of the state's transition to a market-based model for solar project development, the New Jersey Board of Public Utilities (BPU) ordered the state's four regulated electric distribution companies (EDCs) to develop solar financing programs as a means of allowing customers to "securitize" SREC payments. Out of this order, two different program models emerged. Public Service Electric and Gas (PSE&G) modified its pre-existing Solar Loan Program through which the utility loans customers a portion of the up-front cost of a PV system, and the customer repays the loan either in cash or in the form of SRECs generated by their PV system over a 10-15 year term. For purposes of loan repayment, SREC prices are equal to the greater of the prevailing market price for SRECs or a pre-established floor price that varies by customer segment and by loan origination date. PSE&G's current Solar Loan II Program is approved for a two-year period, with a cap of 51 MW. The state's other three EDCs do not directly finance solar projects, but instead, offer 10- to 15-year contracts for the purchase of SRECs, with projects selected through periodic competitive solicitations. In 2009, the BPU approved the contracting programs of these three EDCs for a three-year period, with a cap of 65 MW. The programs offered by all four EDCs are open only to net-metered residential and non-residential systems up to 500 kW.

Importantly, the EDCs are not retail electricity suppliers and therefore do not have RPS compliance obligations. Consequently, the EDCs are allowed to sell the SRECs that they procure to retail electricity suppliers that do have RPS compliance obligation or to other interested parties through an auction process, and use the revenue from such sales to offset the costs of their programs. Effectively, the BPU has used the regulated EDCs to facilitate solar financing through these programs, and therefore to also facilitate compliance with the state's solar set-aside by the state's competitive electricity suppliers.

In addition to these loan and SREC contracting programs, New Jersey has also sought to mitigate SREC price risk and encourage project financing through a variety of other measures. These include: establishing a rolling 15-year solar alternative compliance payment (SACP) schedule and a legislative prohibition on reducing any SACP level that has previously been established in order to provide a measure of certainty to market participants (see Section 3.5 for more on SACPs); and establishing a mechanism for automatically increasing the solar set-aside targets in the event of an SREC surplus and declining SREC prices in three consecutive years, also with the intent of increasing revenue certainty for solar project developers and investors.

Massachusetts: In early 2010, Massachusetts developed emergency regulations for its solar set-aside; as of September 2010, a revision to these rules was under review. Under the emergency rules, owners of eligible solar generators would have the option to participate in an annual *Solar Credit Clearinghouse* auction overseen by the Massachusetts Department of Energy Resources (DOER). A 5% auction fee would be levied on solar system owners, which is intended to encourage project owners to contract bilaterally rather than sell their SRECs through the auction. If project owners do choose to participate in the auction, they must deposit their SRECs into a special auction account. Regular SRECs must generally be sold or retired in the year in which they are generated, but SRECs deposited into the auction account will be retired and re-minted with a two-year life. DOER or its agent will then conduct an auction of the re-minted SRECs. The sale of these special SRECs will be at a fixed price of \$300/MWh (equal to half of the solar alternative compliance payment of \$600/MWh). Bids will be denominated in terms of the *volume* of re-minted SRECs that bidders are willing to buy at the fixed auction price. If an auction does not clear, then the shelf-life of the SRECs is extended to three years, and the auction is repeated.

If the auction again does not clear, then the solar set-aside obligation for the following year is increased and the auction is repeated. In effect, these auctions provide a backstop SREC market – and pricing – for projects that do not arrange bilateral SREC contracts, thereby providing a measure of revenue certainty to solar project developers and investors.

In addition to the auction mechanism, Massachusetts has also sought to encourage SREC price certainty through its approach to setting annual solar set-aside targets.

Each year, the solar target is calculated based on a formula that takes into account the targets in the prior two compliance years, as well as the number of SACPs used, the auction volume, and the number of SRECs banked two compliance years prior. The net effect is that a surplus of SRECs will tend to increase the solar target in subsequent years, while a shortage of SRECs will tend to reduce the target in subsequent years (with the constraint that the target can never be less than in the prior compliance year).

This method for calculating annual solar targets is intended to reduce the likelihood of prolonged periods of depressed or inflated SREC prices, thereby creating greater certainty in SREC prices.

Text Box 3. Emerging Utility Procurement Models for Distributed Solar

Utilities have traditionally sought to add solar energy to their generation mix by offering rebates or other cash incentives for customer-sited PV systems, by purchasing SRECs through short-term over-the-counter markets, or by signing power purchase agreements or long-term REC contracts with utility-scale solar projects.

Increasingly, however, solar/DG set-aside requirements and RPS obligations, more generally, have prompted utilities to develop alternative procurement models – in many cases focusing on distributed solar and often involving utility-ownership of distributed PV assets. The following examples illustrate the diversity of approaches to utility ownership of distributed PV assets that have emerged in recent years:

Public Service Electric & Gas (PSE&G) Solar 4 All Program: This program, launched in 2009, includes two elements, each intended to yield 40 MW of solar capacity by 2013. The first component consists of installing small (approximately 200 watt) PV systems on 200,000 utility poles throughout PSE&G's service territory in New Jersey.

The second component consists of large (>500 kW) rooftop systems and "solar gardens" owned by PSE&G and installed at a combination of PSE&G sites (25 MW), other private property (10 MW), and sites in Urban Enterprise Zones (5 MW).

Arizona Public Service (APS) AZ Sun Program and Community Solar Pilot: APS received regulatory approval in 2010 for two new programs involving utility ownership of distributed PV. Under the AZ Sun Program, APS plans to install 100 MW of utility-owned, ground-mounted PV within its service territory.

Projects will be selected through competitive solicitations and may be located on customer premises, thereby qualifying as distributed energy for compliance with Arizona's RPS DG set-aside. Under the Community Power Project, a pilot program in Flagstaff, APS plans to install 1.5 MW of distributed renewable generation within a single distribution feeder. As part of the program, APS will own, operate, and maintain PV systems installed on residential and commercial customer rooftops, connected on the customer-side of the meter, and will sell the power to the host-customers under longterm power sales agreements.

Duke Energy North Carolina Solar PV Distributed Generation Program: Duke Energy plans to install 10 MW of ground-mounted and rooftop PV systems by 2011 at both residential and non-residential customer sites.

The utility will own the systems, which will be interconnected on the utility-side of the meter, and will provide a monthly lease payment to the customer for the use of their property.

Massachusetts Utility-Owned Solar: The Green Communities Act of 2008 allows the state's regulated Electric Distribution Companies to construct, own, and operate up to 50 MW of solar generation each. To date, two utilities have initiated plans to construct PV systems under the provisions of this law.

National Grid has received pre-approved cost recovery for the construction of 5 MW of PV systems located on company-owned property; later phases of the program will include utility-owned PV projects sited on customer properties as well as financial offerings to customers that want to own a PV system.

Western Massachusetts Electric Company (WMECO) has also received approval for the initial phase of its solar program, in which the utility will develop 6 MW of PV projects at sites that may include both utility properties as well as private and public facilities.

California IOU Programs Targeting Large Distributed PV Projects: Although not subject to a solar set-aside, California's three IOUs – Pacific Gas & Electric (PG&E), Southern California Edison (SCE), and San Diego Gas & Electric (SDG&E) – have received approval for a set of solar PV programs under which the utilities will both own and enter into power purchase agreements (PPAs) for large distributed PV projects. The programs extend over five years, and the three utilities are authorized to procure a total of 1,100 MW of distributed PV (500 MW for PG&E, 500 MW for SCE, and 100 MW for SDG&E).

PG&E's program will target 1-20 MW projects, and the utility-owned projects are expected to be ground-mounted and sited at or near utility substations, while SCE's program will target 1-2 MW projects installed on commercial rooftops, and SDG&E's program will target 1-5 MW, primarily ground-mounted projects. Under PG&E's and SCE's programs, the authorized program capacity is evenly split between utility-owned systems and PPAs, while SDG&E's program is more heavily weighted towards PPAs, representing 74% of the authorized program capacity. These programs are intended to support market segments for solar not adequately supported by other existing programs – namely, the California Solar Initiative, which provides cash incentives for up to 1 MW of capacity, and the general RPS procurement process, in which the transaction costs may effectively exclude participation by smaller distributed PV projects.

3.5. Protocols for SREC Metering, Measurement, Tracking, and Trading

Rules for SREC metering, measurement, tracking, and trading must also be developed in the course of establishing a solar/DG set-aside. A number of states with solar set-asides (Delaware, Maryland, New Mexico, Ohio) rely exclusively on their regional REC tracking systems for these purposes, in which case the same protocols related to REC measurement and tracking apply to solar resources as to other types of renewable generation. Other states, however, have recognized that these protocols may not fit well for small, customer-sited solar facilities and have developed separate rules and procedures, with varying degrees of specificity.

One issue facing RPS administrators is whether to require measurement of the output from small solar electric systems (e.g., residential PV) whose size might not otherwise warrant the expense of high-quality metering hardware.[16]

A number of jurisdictions (Colorado, Missouri, Nevada, New Jersey, North Carolina, Pennsylvania, and Washington D.C.) have therefore adopted metering and measurement requirements for solar projects that vary by system size, and that provide RECs to systems smaller than a specified size threshold (typically 10-15 kW) based on engineering calculations of system output rather than on metered electricity generation. Beyond these states, others with solar or DG set-asides generally require all solar electric systems, regardless of size, to have a dedicated meter for the purpose of determining REC production.[17] In many cases, this latter requirement is implicit, as the RPS rules defer to the measurement procedures developed and applied by the relevant regional REC tracking system, which, in turn, typically require measurement of electricity generation using a revenue-quality meter.[18]

Similarly, states that allow SHC systems to qualify for a solar or DG set-aside must decide what type of metering standards to require for measuring heating and cooling output. A number of jurisdictions (Nevada, North Carolina, and Washington D.C.), under certain circumstances, allow solar thermal output to be estimated, rather than measured directly with a thermal energy meter. Other states require a thermal energy meter for all systems. States that allow SHC systems to qualify for a solar or DG set-aside must also decide how to convert thermal output into electricity (MWh) units. Two such states have specified the conversion within their RPS rules (Arizona and North Carolina specify conversions of 3,415 Btu/kWh and 3,412 Btu/kWh, respectively); other states appear silent on the issue.

Reporting and verification of generation data presents another issue that is addressed by some states. Whereas large generation facilities may be expected to install telemetry to communicate measured output directly to the independent system operator (ISO), utility, or tracking system, this option may be viewed as too costly for smaller systems. States have therefore adopted varying rules related to the reporting of generation data for small systems – including whether meter readings can be self-reported by the system owner or must be conducted by an independent third-party, and how frequently meter data must be reported.[19] Alternatively, states may rely on the protocols established by the regional REC tracking system.[20]

Separate from the issue of how the creation of SRECs is measured and reported, RPS administrators must also establish rules related to the trading and tracking of SRECs. By and large, states with solar or DG set-asides apply the same rules and procedures for tracking and trading SRECs as they do for all other types of RECs.[21] For example, a number of states explicitly allow REC trading (both solar and non-solar), and largely rely upon regional REC tracking systems to ensure that trading in SRECs is appropriately verified; these states include Delaware, Ohio, Pennsylvania, Maryland, New Hampshire, New Mexico, New Jersey, and Washington D.C. New Jersey, however, allows solar projects to qualify for SRECs only for 15 years, after which the projects qualify only for standard New Jersey Class I RECs. Other states have taken somewhat different approaches to developing systems for SREC tracking and trading. In Colorado, for example, utilities may create their own auditable REC tracking database or may request regulatory approval to use a central third-party database. In North Carolina, attestations, contracts, and compliance reports may be used to track REC transactions until regulators have adopted an online tracking platform, expected in 2010, while in Nevada the regulator has developed its own REC tracking database. Similarly, New York bases its SREC tracking on contracts. In New Hampshire, the regional REC tracking system performs most of the tracking functions, but the regulatory body may issue RECs to customer-sited, metered systems located in the state that are otherwise ineligible to participate in that tracking system. In Maryland, the regional REC tracking system also performs most of the tracking functions, however, alternative approaches are allowed for facilities located outside the region.

3.6. Enforcement and Cost Containment

Finally, states employ various approaches to ensure or encourage compliance with solar or DG set-aside targets, while also seeking to contain the absolute cost impact of those policies. As shown in Table 4, many states have established Alternative Compliance Payment (ACP) mechanisms. Though these policies vary in their design, generally electricity suppliers that have been unable to meet their RPS obligations through the purchase of renewable energy or RECs are allowed to make financial payments (ACPs) to meet those obligations. ACPs are distinct from financial penalties, as they are deemed to be legitimate forms of compliance, and suppliers are typically allowed to recover the cost of ACPs from ratepayers, as they would the cost of RECs. The funds collected through ACPs are often used by state agencies to support renewable energy in other ways, although New Jersey has, instead, recently opted to refund solar ACPs to ratepayers. States have typically established higher ACP rates for solar set-asides than for general (non-solar-specific) RPS obligations, reflecting the generally higher cost of solar electricity than other types of renewable energy. Among the six states with solar ACP (SACP) mechanisms in 2009, SACP rates ranged from \$160/MWh to \$711/MWh. Given an expectation that solar electricity costs will decline, however, several states plan to ratchet down SACP rates over time: Maryland and New Jersey have both adopted specified, declining SACP schedules, while Massachusetts' regulators have the discretion to reduce SACP rates by up to 10% per year. In contrast, SACP rates in other jurisdictions (Delaware, Washington D.C., and New Hampshire) either remain constant over time or rise with inflation.

States that have established ACP mechanisms in most cases avoid the need for explicit enforcement procedures in cases of non-compliance with the solar/DG set-aside: receipt of the financial payment fulfills the compliance obligation, and the primary instance of non-compliance would be if an electricity supplier refused to make its ACP. In other states, however, explicit enforcement procedures may be required. Some of these states (Missouri, Ohio, and Pennsylvania) have established financial penalties that are assessed automatically based on the level of non-compliance; in these cases, the penalty is structured similar to an ACP (and may even be called an ACP), except that the fee is not generally recoverable in rates. In other instances (Arizona, Colorado, and Nevada), regulatory commissions have the clear discretion to levy financial penalties (often with unspecified levels) as needed, after notice and hearing. In still other states (New Mexico and North Carolina), the RPS legislation directs the regulatory authority to enforce the targets, but specific rules or procedures have not yet been developed. The level of the SACP can effectively cap the cost of the solar set-aside, because a supplier would presumably not pay more than the SACP to purchase an SREC.[22] The SACP level therefore sets the maximum price for SRECs, and also the maximum cost of compliance for the solar set-aside. Either in lieu of, or in addition to, an ACP, some states have also sought to contain RPS compliance costs by placing a cap on the retail rate increase or per-customer bill increase that is allowable under the RPS (see Table 5). In general, such cost caps are specified only for the entire RPS, in which case the cost of complying with the solar or DG set-aside is counted against the overall cost cap of the RPS policy. However, three states (Delaware, Maryland, and New Jersey) have established retail-rate-based cost caps that are specific to their solar set-asides and separate from the cost caps for their overall RPS. Similarly, New Mexico has specified the maximum price for solar resources allowable under its RPS, which also effectively serves to cap the cost of its solar set-aside. Only four states with a solar or DG set-aside (Arizona, Nevada, New York, and Pennsylvania) do not currently employ an explicit cost cap that has been established through legislation or administrative rules, although the earlier RPS rules in Arizona did contain a cap on the surcharge that utilities could levy on customer bills for recovering RPS compliance costs.

Table 4. Enforcement Provisions for Solar and DG Set-Asides

Alternative Compliance Payments	
DE	SACP for each supplier depends on how many years it has been used: $400/MWhin the first year a supplier uses the ACP; $450/MWh in the second year; $500/MWh in subsequent years
DC	$500/MWh
MA	$600/MWh; regulatory authority has discretion to reduce by up to 10% per Compliance Year
MD	$400/MWh (2009-10); declines by $50/MWh every two years until it reaches$50/MWh in 2023 and then remains at that level
NH	$160/MWh (2009), adjusted by inflation
NJ	$711/MWh (2008-09) declining to $594 (2015-16); 2010 legislation requires BPUto adopt a 15-yearSACP schedule, and to review SACP annually, adding one additional year to the back end of the schedule
Financial Penalties	
MO	Utilities that do not meet their RPS obligations are subject to penalties of at least twice the market value of RECs for the compliance period
OH	Electricity suppliers are charged a pre-specified penalty for all shortfalls in meeting the solar set-aside ($450/MWh in 2009, decliningto $50/MWh in 2024)
PA	Establishes solar compliance fee according to a formula that takes into account solar rebates that have been received; for 2008-2009 (the most recent year posted), that fee was $550.15/MWh
AZ, CO, NV	Financial penalties assessed at the discretion of the regulatory authority
NC, NM	Legislative authority to enforce compliance exists, but no rules have been established that indicate how this will occur

Finally, in addition to cost caps established through legislation or administrative rules, cost containment can also occur through ongoing regulatory oversight of the RPS procurement activities by retail suppliers subject to cost-of-service regulation. Similarly, in New York, the public service commission establishes multi-year program budgets for RPS procurement activities conducted by NYSERDA (the central procurement agent), effectively capping the compliance costs under the state's RPS for the defined, multi-year period.

Table 5. Solar/DG Set-Aside Cost Containment Mechanisms Other than an ACP

Cap on Compliance Costs for *Solar or DG Set-Aside*	
DE	1% increase in retail rates
MD	1% increasein retail rates;allows for one year delay in meeting solar RPS
NJ	2% increasein retail rates
NM	Price of solar cannot exceed $0.15/kWh (projects <10 kW) or $0.10/kWh (projects >10 kW)
Cap on Compliance Costs for *Overall RPS*	
CO	2% increase in retail rates (investor-owned utilities) or1% (cooperatives)
IL	IOUs: Cap on retail rate impact increases each year through 2011, after which it is equal to the greater of 2.015% of average retail rates in 2007 or the incremental RPS costs in 2011; Competitive retail electric suppliers: procurement costs capped based on alternative compliance mechanism23
MO	1% increasein retail rates
NC	Caps on annual cost per account for incremental RPS costs; varies by customer class and year
NM	1% increase in retail rates (2006) rising to 2% (2011); annual dollar caps established for large non-governmental customers ($49,000 in 2006 rising to $99,000 in 2011)
OH	3% increasein generation costs
OR	4% of utility's revenue requirement in the compliance year

Note: This table identifies cost containment mechanisms established through legislation or administrative rule. In addition to the mechanisms identified here, cost containment in many states also occurs through ongoing regulatory oversight of RPS procurement activities and cost recovery, and in New York, cost containment occurs through regulatory approval of multi-year RPS program budgets for the state's central procurement agent (NYSERDA).

4. Solar RPS Policies: Impacts and Expectations

4.1. Overview

Compliance obligations with state solar and DG set-asides have commenced relatively recently. Thus, actual operational experience – both with RPS policies, in general, as well as with RPS policy designs intended to support solar, in particular – still remains rather limited (see Figure 6).

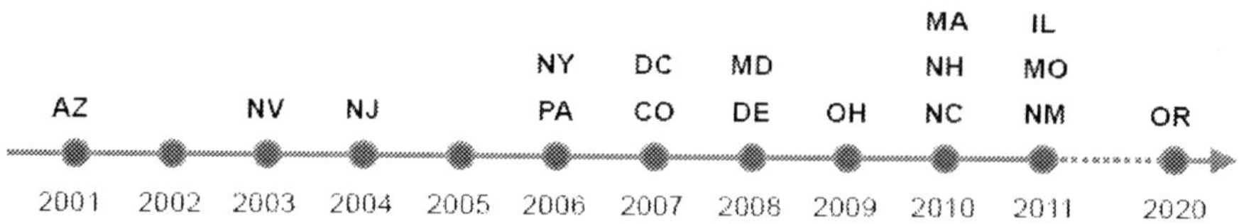

Note: A number of states have RPS compliance years that begin in the middle of the calendar year. For the purpose of this figure, each state is assigned to the first calendar year in which compliance obligations begin.

Figure 6. First Calendar Year with Solar or DG Set-Aside Compliance Obligations.

Nevertheless, early experiences do reveal a number of key trends. Specifically, as discussed at greater length in the pages that follow, while there is little evidence that credit multipliers have succeeded in spurring substantial solar installations, solar and DG set-asides have played a central role in driving growth of solar electric capacity in recent years, especially in certain states (though less so for solar heating and cooling technologies). To date, the retail rate impacts of solar/DG set-asides have been relatively modest, though, for a variety of reasons, set-aside targets have not been universally achieved. Meeting the longer-term solar and DG set-aside targets will require a significant amount of additional solar electric generation capacity in the U.S., signaling that these policies are likely to continue to be important drivers behind growth in the U.S. solar market. At the same time, a spate of recent announcements for utility-scale solar projects suggests that significant growth in solar generating capacity that is *not* associated with solar set-aside requirements may occur throughout the desert southwest and other regions with high quality solar resources. In the remainder of this section, we further explore each of the aforementioned trends.

4.2. RPS-Driven Solar Additions to Date

Solar capacity in the U.S. has expanded rapidly over the past several years, albeit from a relatively small base. This growth has been driven by a number of factors, including (though not limited to) Federal tax incentives, state renewable energy rebate and incentive programs, voluntary green power markets, and RPS programs both with and without solar-specific support mechanisms. Assigning attribution to each of these drivers is challenging.

As one relatively simple indicator of the impact of solar and DG set-asides, in each year from 2005-2009, 65-81% of total grid-connected PV capacity additions in the U.S., outside of California, occurred in states with active or impending solar/DG set-aside obligations (see Figure 7). California, which is by far the largest state PV market in the U.S., does not have a solar set-aside as part of its RPS and has, instead, driven growth in solar capacity through other policy support mechanisms (principally, incentive payments for distributed PV, but more recently though larger-scale solar installations that compete with other forms of renewable energy through the state's traditional RPS, or that receive feed-in tariff payments). In aggregate, from 2000 through 2009, more than 250 $_{MWac}$ of PV capacity is estimated to have been brought online to meet state-level solar or DG set-asides, compared to total U.S. solar capacity additions over that period of approximately of 370 $_{MWac}$ excluding California (or 970 $_{MWac}$ including California). The fact that solar capacity growth outside of California has centered, by and large, in states with solar or DG set-asides, suggests that these policies have played a key role in accelerating solar deployment in the U.S. These impacts have been most apparent in New Jersey (101 $_{MWac}$ through 2009), Colorado (46 MW_{ac}), Arizona (36 MW_{ac}), New York (22 MW_{ac}), and Nevada (19 MW_{ac}); these states were also the largest PV markets in the U.S., after California, from 2007-2009.

The impact of RPS solar set-asides is also evident in the budding resurgence of the CSP market in the U.S. After initial growth in the late 1980's, construction of CSP capacity in the U.S. largely ceased. This has begun to change in recent years, with two new CSP projects constructed to meet solar set-aside requirements: the 1 MW Saguaro project installed in Arizona in 2006 and the 64 MW Solar One facility installed in Nevada in 2007. More recent CSP activity (including three projects completed in 2009 and numerous projects under development) includes many projects that would serve to meet general RPS obligations, as well as solar set-asides.[24]

Unlike solar electric technologies, growth in solar heating and cooling capacity thus far has not been strongly influenced by the presence of state RPS programs, in general, or solar/DG set-asides, more specifically. This is, in part, a consequence of the fact that SHC technologies are eligible for only a sub-set of RPS set-asides (as shown previously in Table 3), and even in those cases, the types of eligible applications may be limited (e.g., to only applications that displace electricity consumption). The top three states in terms of SHC capacity additions – constituting 50-56% of SHC-related solar thermal collector shipments each year from 2006-2008 – are Hawaii, California, and Florida, none of which have a solar/DG set-aside (Sherwood 2010). Excluding SHC capacity additions in these states, still just 13% of the remaining solar thermal collector shipments in 2008 (or 6% of the U.S. total) occurred in states where SHC was eligible for a solar set-aside then in effect.

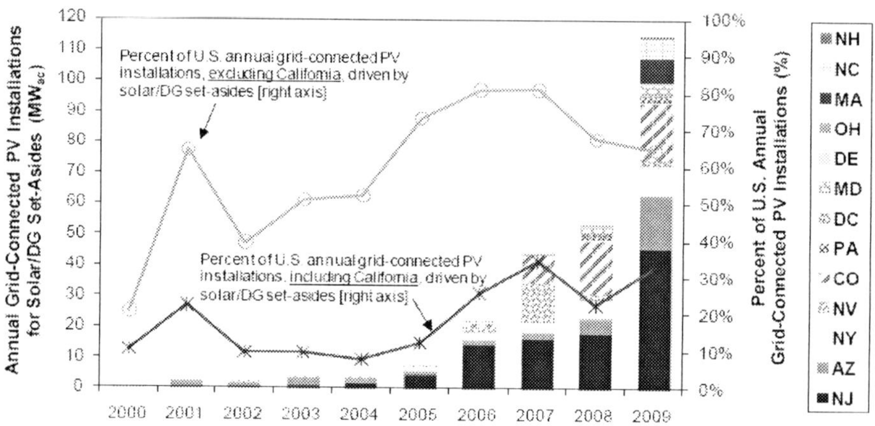

Notes: For the purpose of constructing this figure, solar capacity additions were identified as being driven by solar or DG set-asides if and only if they are located in a state with a set-aside and began operation no sooner than one-year before the start of compliance obligations under the state's set-aside. The only exception is the 10 MW_{ac} El Dorado PV project installed in Nevada in 2008; the electricity generated by this project is being sold into California, and therefore is not attributed to Nevada's solar set-aside.

Figure 7. Grid-Connected PV Capacity Additions Driven by Solar and DG Set-Asides.

Although experience with *credit multipliers* targeting solar energy is extremely limited, the available evidence suggests that, at least to date, they have not resulted in a significant increase in solar generation. The Colorado RPS, for example, includes a solar credit multiplier for the state's rural electric cooperatives and municipal utilities, where each kWh of solar counts as three kWh for RPS compliance. Through 2008 (the most recent year for which RPS compliance data are available), Colorado's municipal utilities and cooperatives fully achieved their RPS compliance obligations with virtually no contribution from solar energy resources. In contrast, the state's investor-owned utilities are subject to a solar set-aside that has resulted in a significant growth in solar capacity. Three other states (Washington, Texas, and Michigan) currently have

RPS policies that include only a credit multiplier for solar without an accompanying set-aside. RPS obligations have not yet begun in Washington and Michigan, thus it remains to be seen whether the solar credit multipliers in those states succeed in stimulating solar development. In Texas, the non-wind multiplier (and associated set-aside goal) may have had a limited impact on solar development to date, although Texas' final RPS compliance obligations were fully achieved in 2008, obviating the need for any further renewable additions (solar or otherwise) for meeting its RPS. In the past, both Maryland and Delaware offered RPS credit multipliers for solar energy without an accompanying set-aside. During the brief windows of time when compliance obligations existed in each state but only credit multipliers were in force, relatively small amounts of solar were installed (200-400 kW per year in Delaware, and <100 kW per year in Maryland). Finally, in Nevada, which currently has a multiplier for solar PV, along with a broader solar electric set-aside, the PV multiplier has had some effect in steering solar development towards PV and away from CSP (though both PV and CSP development have occurred under the state's RPS).

4.3. Compliance with RPS Solar Set-Asides

Notwithstanding the general success of solar/DG set-asides in expanding solar markets in the U.S., "compliance" with these requirements has been mixed. As of 2008 (the most recent year for which comprehensive solar set-aside compliance data is available), nine states had active solar or DG set-aside compliance obligations. Across these nine states, 68% of the aggregate solar/DG compliance obligation in 2008 was achieved through the purchase of solar energy, DG, and/or SRECs (see Table 6). Though the requirements in a number of these states amounted to only several MWs of solar capacity, only three states fully achieved their solar/DG requirements in 2008; the other six states met from 0% to 84% of the solar/DG set-aside obligations with the purchase of eligible forms of renewable energy. Though electricity suppliers in these states may not have been technically out of compliance with the solar/DG set-asides (due to SACP compliance options, funding limits, and force majeure provisions), these results demonstrate the challenges of meeting even early-year solar/DG set asides.

A time trend showing "compliance" results just for those states that had relatively sizable solar/DG requirements through 2008 is presented in Figure 8. In Arizona, the state with the longest history of compliance obligations under a solar or DG set-aside, compliance has remained below 40% since 2002, even after accounting for credit multipliers. This has largely been the result of RPS funding caps in place under the previous set of RPS rules, which were below the level necessary to meet the state's RPS targets.[25] In New Jersey, the solar set-aside targets were almost fully achieved in 2005 and 2006.

Table 6. 2008 Solar/DG Set-Aside Compliance Results

State	2008 Solar/DG Set-Aside Compliance Obligation		Percent of "Compliance" Obligation Achieved*
	% of Applicable Retail Sales	Equivalent Capacity @ 15% capacity factor (MWac)	
Nevada	0.54%	104	100%
New Jersey	0.16%	99	58%
New York	0.07%	58	27%
Arizona	0.18%	52	40%
Colorado	0.20%	46	100%
Maryland	0.01%	2	7%
Washington D.C.	0.01%	1	0%
Pennsylvania	0.01%	1	100%
Delaware	0.01%	1	84%
		Weighted Average	68%

* Percent of "Compliance" Obligation Achieved excludes ACPs but includes applicable credit multipliers. In cases where this figure is below 100%, suppliers may not have been technically out of compliance due to solar ACP compliance options, funding limits, and force majeure provisions.

Since that time, however, SREC supply has not kept pace with the annually increasing targets, resulting in a growing portion of the obligation being met through SACPs; in 2008, approximately 60% of the total set-aside obligation was met through retirement of SRECs, with the remaining portion met with SACPs. Despite various efforts to encourage solar project financing by the state's policymakers and lucrative SREC pricing, long-term SREC contracts are not widely available from competitive electricity suppliers in New Jersey, slowing solar development and thus far impeding full compliance absent heavy reliance on the SACP. Compliance with Nevada's solar set-aside, on the other hand, has steadily risen in recent years after two large solar projects – the 14 MW PV facility at Nellis Air Force Base and the 64 MW Nevada *Solar One* CSP plant – came online in 2007, providing the bulk of the solar generation required to meet the Nevada utilities' solar set-aside requirements for several years to come.[26] Colorado also fully met its solar requirements in 2008, relying primarily on SRECs purchased through its customer-sited PV incentive program, along with the 8 MW Alamosa PV project. New York, in contrast, has fallen well-short of the DG required under the Customer-Sited Tier of its RPS program, with eligible projects installed through 2008 representing just 27% of the target. Similar to Arizona, the funding levels established for meeting New York's DG requirements were insufficient for achieving its targets.

The remaining four states shown in Table 6 (Delaware, Maryland, Pennsylvania, and Washington D.C.) all had relatively small solar obligations through 2008, in terms of the absolute amount of solar generation required – in each case, requiring less than the equivalent of 2 MW of solar capacity. Of these four states, Maryland and Washington D.C. both missed their solar set-aside targets more or less in their entirety in 2008, reflecting the nascent state of the PV market in those jurisdictions as well as the challenges of encouraging long-term contracting and facilitating solar project financing in markets open to retail electricity competition. In contrast, the small solar targets in Pennsylvania and Delaware were fully or largely achieved.

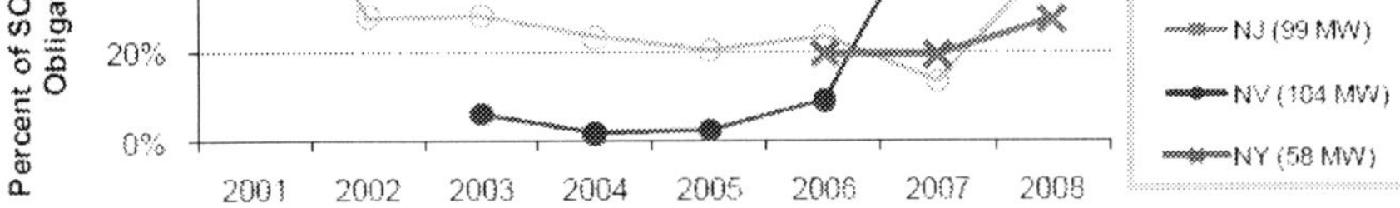

Note: Percent of "Compliance" Obligation Achieved excludes ACPs but includes applicable credit multipliers. In cases where this figure is below 100%, suppliers may not have been technically out of compliance due to solar ACP compliance options, funding limits, and force majeure provisions.

Figure 8. Compliance with Solar/DG Set-Aside Targets over Time.

4.4. SREC Prices and Retail Rate Impacts

Most states with solar/DG set-asides allow unbundled SRECs to be used for compliance. In some cases, SRECs are transacted solely through long-term bilateral contracts, for which prices typically are not publicly available. In a number of states, however, SRECs are (also) traded through short-term spot markets, providing some visibility into both the value of potential investments in solar as well as the overall cost to electricity consumers (i.e., the retail rate impacts) of the solar/DG set-asides.

As evident in Figure 9, SREC markets in the U.S. remained somewhat fragmented through 2009, reflecting differences in the underlying policies and the fact that some states prohibit or limit the use of SRECs sourced from out-of-state or out-of-region facilities.[27] In general, SREC prices were highest in those states with the highest SACPs (as identified in the legend of Figure 9), though average monthly SREC prices generally remained below SACP levels in each state. The similar SREC prices observed among Delaware, Maryland, Ohio, and Pennsylvania may partly reflect the similar SACP levels across those states, as well as the opportunity to trade SRECs among most of these states (as described previously in Table 3).

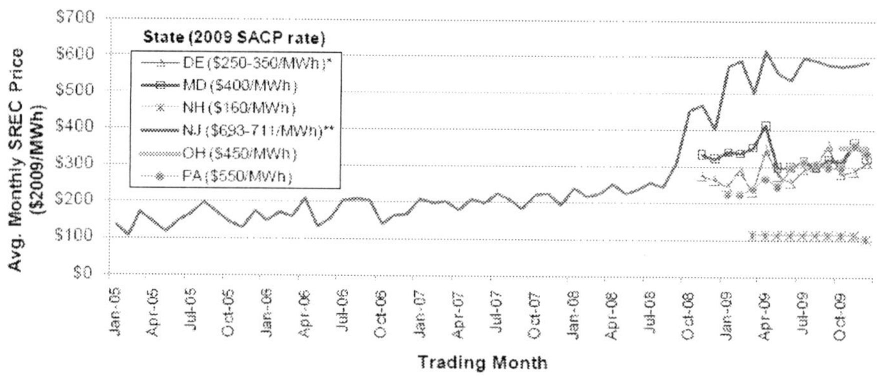

Sources: New Jersey Clean Energy Program (NJ), Spectron (NH), PJM-GATS (all other states). Plotted values are the weighted average selling price, except NH, where they are the mid-point of the reported Bid and Offer prices for the current or nearest compliance year.

* Delaware's SACP rate varies by supplier, depending on how many years the supplier has availed itself of the SACP option. Under the RPS rules in place in 2009, the SACP rate ranged from $250-350/MWh. New legislation passed in 2010 increased Delaware's SACP rates to the levels described in Table 4.

** New Jersey's SACP rate was $711/MWh during the first five months of 2009 and was $693/MWh for the latter seven months.

Figure 9. Solar Renewable Energy Credit Spot Market Prices.

The figure also illustrates the potential for significant fluctuations in SREC prices within an individual state. In particular, SREC prices in New Jersey rose substantially in 2008 and 2009, as the solar targets ratcheted up and SACP rates increased from $300/MWh to $711/MWh in mid-2008. In other states, all of which have a much shorter SREC price history, such fluctuations have not been observed.

The cost of complying with RPS solar/DG set-aside requirements has not been compiled in a comprehensive fashion, in part because of the early status of policy implementation and in part because of methodological complexities and data availability constraints.

Table 7. Estimated Retail Rate Impact of Solar/DG Set-Asides in 2009

State	Solar/DG Target (% of retail sales)	Retail Rate Impact (% of total retail costs)
Arizona	0.30%	1.15%*
Delaware	0.01%	0.03%
Maryland	0.01%	0.04%
New Jersey	0.20%	0.96%
New York	0.10%	0.01%*
Ohio	0.004%	0.04%
Pennsylvania	0.01%	0.04%
Washington D.C.	0.02%	Unknown (data not available)
Colorado	0.20%	
Nevada	0.72%	

* The estimated retail rate impacts shown for Arizona and New York are based on actual/budged expenditures for DG incentive programs, which we take to be a proxy for the incremental cost of DG resources, though not strictly comparable to SREC prices.

The SREC prices presented in Figure 9 can, however, be used to roughly estimate the cost of complying with solar set-aside requirements in those states *if* one assumes: (a) that SREC prices represent the total incremental above-market cost of solar resources, (b) that the short-term SREC prices presented in the figure are representative of all SRECs used for set-aside compliance, and (c) that the full compliance obligation was achieved solely through retirement of SRECs, without the need for SACPs, or that SACP and SREC prices are similar.[28] Given these assumptions, the estimated average increase in retail electricity prices associated with meeting solar/DG set-aside requirements in 2009 is shown in Table 7, for the five states for which short-term SREC prices are available[29], along with Arizona and New York, where estimated retail rate impacts are based on actual or budgeted funding levels for DG incentive programs.[30] For five of these states (Delaware, Maryland, New York, Ohio, and Pennsylvania), the rate impacts are estimated at 0.04% or less, reflecting the correspondingly low solar targets. In New Jersey and Arizona, where the set-aside targets in 2009 were appreciably greater, estimated compliance costs reached 0.96% and 1.15% of total retail electricity costs, respectively.

4.5. Projected Solar Capacity Required to Achieve Set-Aside Targets

The impacts of RPS solar/DG set-asides on solar development will continue to grow as a greater number of the existing set-asides take effect and as targets increase over time. By 2025, we estimate that approximately 9,500 MWac of solar generation capacity would be needed to fully meet the existing state set-aside requirements (see Figure 10).[31] Four states (New Jersey, Illinois, Arizona, and Maryland) represent the bulk of this total. Fully meeting RPS set-aside targets during intervening years would require average annual capacity additions on the order of 400 MWac per year from 2010-2014 and approximately 600 MWac per year from 2015-2025. In comparison, in 2009, approximately 107 MWac of solar capacity was added in order to meet solar and DG set-asides in the U.S., a figure that would have been higher had full compliance with existing targets been achieved.

Meeting solar set-aside targets in the future will therefore require a scale-up in the solar delivery infrastructure in many U.S. states. In addition to this challenge, a number of policy design issues may constrain the market's growth to levels below those noted above. In particular, as described earlier in Table 5, many states have developed cost containment mechanisms that may ultimately become binding, thereby limiting future solar capacity additions to levels below those estimated here. For example, Arizona and New York both missed their set-aside targets in 2008 as a result of funding limits, and Ohio utilities were granted a force majeure exemption in 2009, thereby curtailing achievement of that state's solar targets. Furthermore, some states, especially those with competitive retail electricity markets, continue to struggle with how to encourage long-term contracting for solar generation. In part as a result, compliance with solar set-aside targets to date has been well below 100%, in aggregate.

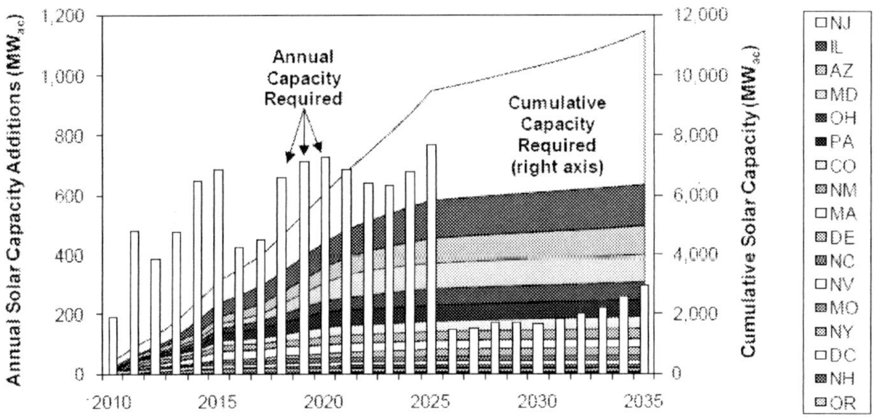

Figure 10. Required Solar Capacity Additions to Meet Solar/DG RPS Set-Aside Targets.

Solar and DG set-asides, combined in many cases with additional financial incentives, have been the primary force behind solar electric capacity additions outside of California in recent years, and as shown in Figure 10 will continue to motivate capacity growth going forward. That said, recent announcements for utility-scale solar projects in the U.S. suggest that future growth in solar capacity will, by no means, be limited to meeting set-aside obligations (see Table 8). Among the approximately 23,000 MW of announced utility-scale solar projects in the U.S., more than 15,000 MW is planned for California, where (if constructed) these projects would presumably serve the state's general RPS obligations. Outside of California, most of the remaining announced capacity is planned for states in the desert southwest (Arizona, Nevada, and New Mexico). Some portion of that announced capacity would likely serve the solar set-aside obligations in those states. However, much of the solar capacity planned in Arizona consists of central-station projects, which are not eligible for that state's DG set-aside, and the planned solar capacity in Nevada and New Mexico significantly exceeds those states' solar set-aside requirements through 2020. Thus, much of the announced solar capacity in the desert southwest would, if built, likely end up serving general RPS obligations in the host states or other states in the region, rather than solar set-aside targets specifically. Outside of the southwest and California, utility-scale solar projects have been announced in states with solar set-asides (380 MW) as well as in states with only general RPS obligations (97 MW). In addition, a sizable amount of solar capacity has been announced in Texas (381 MW), where the RPS targets have already been fully met, and in Florida (208

MW), which does not have an RPS. Florida has already seen the addition of one utility-scale solar facility, the 25 $_{MWac}$ DeSoto plant, which was completed in 2009 and represented the largest PV plant in the U.S. at the time of commercial operation.

Collectively, these project announcements suggest that, at least in the desert southwest and other regions with relatively strong solar resources, utility-scale solar may be able to compete against other renewables within a general RPS framework. As a result, in the future, state solar/DG set-asides may be especially important in states with poorer solar resources or for solar applications that are not likely to fare well under a traditional RPS design (e.g., customer-sited, distributed solar).

Table 8. Summary of Announced Utility-Scale PV and CSP Projects

State	MW	RPS Drivers
CA	15,492	General RPS obligations
NV	3,367	Solar set-aside (~110 MW target in 2020) and general RPSobligations
AZ	2,225	General RPS obligations (primarily) and DG set-aside
NM	429	Solar set-aside (~300 MW target in 2020) and general RPSobligations
TX	381	Not RPS-driven
FL	208	Not RPS-driven
Other states with solar/DG set-asides (CO, DE, MA, MD, NC, NJ, NY, OR, PA)	380	Solar set-aside and/or general RPS obligations
Other RPS states without solar/DG set-aside (HI, MN, WA)	97	General RPS obligations
Other states without RPS (ID, GA, TN, VT)	38	Not RPS-driven
Total	22,617	

Source: LBNL analysis of data compiled by the Solar Energy Industries Association.

CONCLUSIONS AND LESSONS LEARNED

By design, most RPS policies were originally developed to be largely technology-neutral, stimulating competition among various renewable energy technologies and allowing the most economically attractive technologies to win out. Not surprisingly then, experience has generally shown that traditional

RPS programs in which all eligible renewable technologies compete have yielded rather modest levels of renewable resource diversity. This realization has led many U.S. states to design their RPS policies to provide targeted support for solar (and other preferred) renewable technologies. Although most of these RPS provisions have been in place for only a brief number of years, the breadth of experience and variation in policy design reveals a number of key findings.

RPS programs can spur the development of solar energy. Perhaps the most important conclusion is simply that RPS programs can, in some cases, lead to the development of solar energy and to renewable resource diversity, more generally. This fact is evident by the strong growth in solar electric capacity and the pipeline of solar projects under development in California, New Jersey, Nevada, Colorado, New York, Arizona, and other states. In most of these states, this market growth has been driven largely by RPS policies with provisions that specifically target solar or DG. However, the ability of RPS programs to successfully achieve resource diversity, and to do so at an acceptable cost, requires numerous tradeoffs in policy design, and careful attention to policy details.

Set-asides have been a more popular, and arguably more effective, RPS mechanism for solar support than credit multipliers. Of the two basic mechanisms for providing targeted support to solar technologies through an RPS – solar set-asides and solar credit multipliers – set-asides have clearly emerged as the more popular option. This preference is likely driven by a number of perceived advantages of solar set-asides, compared to credit multipliers. Most obvious, perhaps, is that set-asides provide greater certainty that a given amount of solar energy will be produced. This presumption has, in fact, been born out in experience, as solar set-asides have successfully initiated solar market growth in a number of states, while solar credit multipliers have yet to demonstrate any comparable success (though this may partly be due to a shortage of actual operating experience with solar credit multipliers).

Targeted forms and degrees of support may be warranted for different types of solar applications and project sizes. Solar is somewhat unique among renewable energy technologies in the extent to which the technology can serve a variety of distinct applications and relatively localized markets. RPS programs may therefore require tailored strategies focused on different solar market segments in order to achieve broad deployment among multiple applications and meet aggressive set-aside targets. For example, the recent spate of announcements for utility-scale solar projects throughout the southwest (including California) and Florida suggests that, at least in regions with strong solar resources, utility-scale solar projects may be competitive with wind power and other forms of renewable generation. Some states may therefore wish to consider whether solar set-asides continue to be appropriate for utility-scale solar projects, or whether such policies are best targeted to distributed generation applications. Furthermore, within the distributed PV market, project development in many regions has become increasingly oriented towards relatively large-scale distributed PV applications (e.g., >100 kW installations), due in part to economies of scale. State policymakers may therefore also wish to consider whether differentially greater support for small-scale distributed PV projects is warranted – e.g., a requirement that some portion of the solar set-aside be met with residential projects, as well as mechanisms to make it easier for small project owners to participate in SREC markets.

The success of RPS policies in fostering the development of solar energy hinges on whether financing-related barriers are adequately addressed. Experience with RPS policies has often revealed the challenges faced by renewable projects in securing financing, especially in restructured markets where RPS compliance has tended to be dominated by short-term REC transactions. For solar energy, these issues can be particularly acute, given the smaller project sizes, higher up-front costs, and – at least until recently – the less-mature state of commercial development of solar relative to certain other renewable technologies (e.g., wind power). As a prime illustration of their role as "laboratories of experimentation," several states have introduced innovative mechanisms for mitigating this revenue uncertainty. In New Jersey, regulated distribution utilities hold auctions for long-term SREC contracts and re-sell those SRECs to the competitive electricity suppliers that must meet the state's solar set-aside. In Massachusetts, the state plans to hold annual statewide auctions for SRECs, and has established a floor price that will remain in effect over the lifetime of the program. And in New York and Illinois, state agencies play an active role in procuring RECs required for RPS compliance. Further experimentation with such mechanisms may be needed to help ensure that solar set-aside targets are fully achieved in a least-cost fashion, especially in markets open to retail electricity competition.

Solar set-asides may complicate RPS cost containment. Although relatively few states have thus far reached their cap, costs caps are likely to increasingly become an issue as RPS compliance obligations rise over time. Already, funding limits in both Arizona and New York have been binding, resulting in underachievement of each state's solar or DG set-aside target. Solar set-asides may also put pressure on general RPS cost caps. Solar energy has historically been more expensive than those forms of renewable energy that have traditionally been used to meet RPS compliance obligations (principally wind power). To the extent that these historical trends persist, solar set-asides may raise RPS compliance costs. In states where RPS cost containment is achieved solely through ACP mechanisms, the higher cost of solar resources has been accommodated by establishing a separate and higher ACP rate for solar resources than for general RPS compliance. States with explicit cost or rate impact caps, however, have often established only overall RPS caps, without a separate cap for the solar set-aside. These states may find that their RPS cost cap is reached sooner than originally anticipated, resulting in less renewable energy development. States with cost or rate impact caps may therefore wish to consider either increasing the overall RPS cap or establishing separate caps for their solar and DG set-asides.

REFERENCES

Agnolucci, P. 2007. "The importance and the policy impacts of post-contractual opportunism and competition in the English and Welsh non-fossil fuel obligation." *Energy Policy*. 35(1): 475-486.

Bergek, A. and S. Jacobsson. 2010. "Are tradable green certificates a cost-efficient policy driving technical change or a rent-generating machine? Lessons from Sweden 2003–2008." *Energy Policy*. 38(3): 1255-1271.

Butler, L. and K. Neuhoff. 2008. "Comparison of feed-in tariff, quota and auction mechanisms to support wind power development." *Renewable Energy*. 33(8): 1854–1867.

CEC (California Energy Commission). 2010. Database of POU Contracts for Renewable Generation: September 2010 Update. http://www.energy.ca.gov/2008publications/CEC-300-2008-005/CEC-300-2008-005_rev.xls (accessed September 23, 2010).

Chupka, M. 2003. "Designing effective renewable markets." *The Electricity Journal*. May: 46- 57.

CPUC (California Public Utilities Commission). 2010. "RPS Project Status Table – September 2010 Update." San Francisco, CA: California Public Utilities Commission. http://www.cpuc.ca.gov/PUC/energy /Renewables /index.htm (accessed September 23, 2010).

Cory, K. and B. Swezey. 2007. "Renewable portfolio standards in the states: Balancing goals and rules." *The Electricity Journal*. 20(4): 21-32.

Endrud, N. 2008. "State renewable portfolio standards: Their continued validity and relevance in light of the dormant commerce clause, the supremacy clause, and possible federal legislation." *Harvard Journal on Legislation*. 45: 259-286.

Exeter Associates, Inc. 2008. *Progress Report: Review of State Renewable Portfolio Standard Programs in the Northeast and Mid-Atlantic Regions*. Montpelier, Vermont: Clean Energy States Alliance.

Falconett, I. and K. Nagasaka. 2010. "Comparative analysis of support mechanisms for renewable energy technologies using probability distributions." *Renewable Energy*. 35(6): 1135-1144.

Ferrey, S. 2006. "Renewable orphans: Adopting legal renewable standards at the state level." *The Electricity Journal*. 19(2): 52-61.

Finon, D. and Y. Perez. 2007. "The social efficiency of instruments of promotion of renewable Eenergies: A transaction-cost perspective." *Ecological Economics*. 62(1): 77–92.

Haddad, B and P. Jefferiss. 1999. "Forging consensus on national renewables policy: The renewables portfolio standard and the national public benefits trust fund." *The Electricity Journal*. March: 68-80.

Holt, E. and R. Wiser. 2007. "The Treatment of Renewable Energy Certificates, Emissions Allowances, and Green Power Programs in State Renewables Portfolio Standards." Berkeley, CA: Lawrence Berkeley National Lab. LBNL-62574.

Hvelplund, F. 2001. *Renewable Energy Governance Systems: A comparison of the "political price-/amount market" model with the "political quota-/certificate price market" system (The German and Danish cases)*. Aalborg, Denmark: Aalborg University. ISBN 87-90893-26-3.

IEA (International Energy Agency). 2008. *Deploying Renewables - Principles for Effective Policies*. Paris, France: International Energy Agency. ISBN 978-92-64-04220-9.

Keydes, A. 2007. "Impacts of a renewable portfolio generation standard on US energy markets." *Energy Policy*. 35: 809-814.

Kildegaard. A. 2008 "Green certificate markets, the risk of over-investment, and the role of longterm contracts." *Energy Policy*. 36(9): 3413-3421.

Lauber, V. 2004. "REFIT and RPS: options for a harmonised Community framework." *Energy Policy*. 32(12): 1405-1414.

Lipp, J. 2007. "Lessons for effective renewable electricity policy from Denmark, Germany and the United Kingdom." *Energy Policy*. 35(11): 5481-5495.

Menanteau, P., D. Finon and M. Lamy. 2003. "Prices versus quantities: choosing policies for promoting the development of renewable energy." *Energy Policy*. 31(8): 799-812.

Meyer, N. 2003. "European schemes for promoting renewables in liberalised markets." *Energy Policy*. 31(7): 665-676.

Meyer, N. and A. Koefoed. 2003. "Danish energy reform: policy implications for renewables." *Energy Policy*. 31(7): 597-607.

Michaels, R. 2008. "A national renewable portfolio standard: Politically correct, economically suspect. *The Electricity Journal*. 21(3): 9-28.

Midttun, A. and K. Gautesen. 2007. "Feed in or certificates, competition or complementarity? Combining a static efficiency and a dynamic innovation perspective on the greening of the energy industry." *Energy Policy*. 35(3): 1419–1422.

Mitchell, C., D. Bauknecht and P.M. Connor. 2006. "Effectiveness through risk reduction: a comparison of the renewable obligation in England and Wales and the feed-in system in Germany." *Energy Policy*. 34(3): 297-305.

Nogee, A., J. Deyette and S. Clemmer. 2007. "The projected impacts of a national renewable portfolio standard." *The Electricity Journal*. 20(4): 33-47.

Palmer, K. and D. Burtraw. 2005. "Cost-effectiveness of renewable electricity policies." *Energy Economics*. 27: 873-894.

Rabe, B. 2006. *Race to the Top: The Expanding Role of U.S. State Renewable Portfolio Standards*. Washington, DC: Pew Center for Global Climate Change.

Rader, N. and R. Norgaard. 1996. "Efficiency and sustainability in restructured electricity markets: A renewable portfolio standard." *The Electricity Journal*. 9(6): 37-49.

Ringel, M. 2006. "Fostering the use of renewable energies in the European Union: the race between feed-in tariffs and green certificates." *Renewable Energy*. 31(1): 1–17.

Sherwood, L. 2010. Personal communication providing data on solar hot water collector shipments by state. March 9, 2010.

van der Linden, N.H., M.A. Uyterlinde, C. Vrolijk, L.J. Nilsson, J. Khan, K. Åstrand, K. Ericsson and R. Wiser. 2005. *Review of International Experience with Renewable Energy Obligation Support Mechanisms*. Energy Research Centre of the Netherlands. ECN-C--05- 025.

Verbruggen, A. 2004. "Tradable green certificates in Flanders (Belgium)." *Energy Policy*. 32: 165-176.

Voogt, M.H. and M.A. Uyterlinde. 2006. "Cost effects of international trade in meeting EU renewable electricity targets. *Energy Policy*. 34(3): 352-364.

Wiser, R., K. Porter, M. Bolinger and H. Raitt. 2005a. "Does it have to be this hard? Implementing the nation's most complex renewables portfolio standard." *The Electricity Journal*. 18(8): 55-67.

Wiser, R., K. Porter and R. Grace. 2005b. "Evaluating experience with renewable portfolio standards in the United States." *Mitigation and Adaptation Strategies for Global Change*. 10: 237-263.

Wiser, R. and G. Barbose. 2008. *Renewables Portfolio Standards in the United States: A Status Report with Data Through 2007*. Berkeley, CA: Lawrence Berkeley National Lab. LBNL154E.

Yin, H. and N. Powers. 2010. "Do state renewable portfolio standards promote in-state renewable generation?" *Energy Policy*. 38: 1140-1149.

End Notes

[1] California, which is by far the largest state PV market in the U.S., does not have a solar set-aside as part of its RPS and has, instead, driven growth in solar capacity through other policy support mechanisms: principally, through incentive payments for distributed PV, but more recently though larger-scale solar installations that compete with other forms of renewable energy within the state's RPS or that receive feed-in tariff payments.

[2] Data on spot market SREC prices can be used to roughly estimate the cost of complying with solar set-aside requirements in those states *if* one assumes: (a) that SREC prices represent the total incremental above-market cost of solar resources, (b) that the average price of short-term SRECs is representative of the price of all SRECs used for compliance, and (c) that the full compliance obligation was achieved solely through retirement of SRECs, without the need for solar alternative compliance payments (SACPs), or that SREC and SACP prices are similar.

[3] Such mechanisms have been discussed in the literature for some time (see, e.g., Rader and Norgaard 1996; Haddad and Jefferiss 1999; and Verbruggen 2004), but few evaluations of their design and effectiveness have yet been published.

[4] For additional and more-general information on U.S. state RPS policies and experiences, not only those focused on solar resources, see: Rabe (2006), Exeter (2008), and Wiser and Barbose (2008).

[5] At the Federal level, the U.S. House of Representatives and Senate have, at different times, each passed versions of a national RPS, although none has thus far been signed into law.

[6] Using a more sophisticated approach, Yin and Powers (2010) find that U.S. state RPS programs have had a statistically significant and positive impact on in-state renewable energy development.

[7] Because each unit of renewable energy on which a multiplier applies counts as more than a single MWh towards RPS compliance, higher multipliers will (absent a chance in the underling renewable energy target) naturally lead to lower levels of renewable energy deployment.

[8] In addition to being used to support resource diversity, set-asides and multipliers have also been used within RPS programs to, among other things, support the development of in-state renewable resources.

[9] Because of various exemptions offered to certain electricity suppliers or customers, RPS obligations do not always apply to all retail sales in a state (Wiser and Barbose 2008),

[10] The percentage targets shown in Figure 5 are translated into solar capacity additions (MW) later, in Figure 10.

[11] Solar heating and cooling applications may offset the onsite use of natural gas or oil rather than electricity, which would offer greater scope for the reduction of greenhouse gas emissions; however, inclusion of SHC within an RPS may be problematic if the state's legal authority over the RPS stems from electricity regulation. Perhaps as a result, a number of states that allow SHC limit eligible applications to those that displace electricity use.

[12] This very issue was recently brought to the fore through a lawsuit filed by TransCanada Power Marketing in U.S. District Court against the state of Massachusetts. The lawsuit challenged the in-state provisions of two separate state renewable energy programs: the RPS solar set-aside and a separate program mandating that the state's distribution utilities solicit long-term contracts with renewable generators located either in-state or in adjacent Federal waters. Under the settlement reached by the parties, Massachusetts retained the in-state requirement under its RPS solar set-aside, but eliminated this requirement under the long-term contracting program. For more on the interstate commerce clause and its implications for the legality of state renewable energy policies, see Ferrey (2006) and Endrud (2008).

[13] Though no longer in existence for new facilities, Arizona's RPS once contained an extensive and varied list of multipliers, and Washington D.C. employed a time-limited multiplier for new solar that has since expired.

[14] Colorado's multiplier for solar energy applies for a limited time to the state's POUs, which are not obligated to meet the DG set-aside that applies to the state's IOUs.

[15] In Colorado, the utility regulator requires that IOUs offer a standard rebate of $2/W for customer-sited PV systems; in addition to the standard rebate, Xcel Energy offers a separate payment for the SRECs produced by such systems over a 20-year period. The Missouri RPS legislation obligates electric utilities to offer a rebate of $2/watt for customer-sited solar systems up to 100 kW, but the rebate is to be paid for only up to 25 kW. Subsequent regulations adopted by the Missouri Public Service Commission require the utilities to also develop standard-offer contracts that provide either an up-front or annual payments for solar RECs produced over 10 years, with terms that vary depending on project size. Similarly, Nevada requires that each utility in that state offer up-front, declining rebates to certain solar systems, with total payment based on the lesser of the actual capacity of the system or 5 kW for residential systems, 30 kW for public buildings and small business systems, and 50 kW for school systems. By accepting the rebate, a program participant is required to transfer the SRECs to the utility. Alternatively, a participant may opt to enter into an agreement with the utility to transfer the SRECs over time for a minimum term of ten years. In New York, as part of its RPS, a state entity (the New York State Energy Research and Development Authority, NYSERDA) offers rebates to solar installations based on system size, differentiated by sector. Finally, as part of a law that increased its RPS in 2008, Washington, D.C. also adopted a renewable energy incentive program that includes up-front rebates for small solar PV and SHC projects.

[16] Most small systems are net-metered, requiring only one bi-directional meter and resulting in the measurement of the net excess generation (beyond any solar used onsite), if any, fed from the customer's system into the distribution grid. This limited measurement does not

record the entire generation produced by the system, most of which is consumed onsite, so cannot be used as a means of issuing RECs to the owner of the system.

[17] Where separate metering of solar energy systems is required in order to assign RECs, this raises the issue of which party (the customer or the utility) will bear the cost of the metering equipment. Colorado and North Carolina both assign that cost to the utility. Other states generally are not clear about who pays for the meter, although many states' net metering rules prohibit a utility from compelling a customer to pay for a second meter.

[18] The regional REC tracking systems for New England, the Mid-Atlantic, the West and the Upper Midwest all require, by default, measurement with a revenue-quality meter. However, the tracking system in the Mid-Atlantic region allows generation from small solar systems to be estimated using the "PVWatts" software developed by the National Renewable Energy Laboratory, if permitted by state regulations.

[19] For example, New York allows generation data from all solar energy systems to be self-reported, while New Hampshire requires that it be performed by an independent third-party, and North Carolina allows self-reported data for systems under 1 MW. In New York, meters must be read at least every six months and reported to NYSERDA at least twice a year for three years. For systems 25 kW and larger, meter readings must be taken monthly and submitted to NYSERDA every six months. New Hampshire requires a meter reading no less frequently than 13 months (the first within 30 days of the beginning of the calendar year and the second within 30 days of the end of the year); the District of Columbia requires a meter read at least once each calendar year.

[20] Most regional REC tracking systems allow meter readings from small generators to be reported by the owner of the generation unit on either a quarterly or annual basis.

[21] See Holt and Wiser (2007) for a more comprehensive summary of state RPS rules related to REC trading and tracking.

[22] Financial penalties, on the other hand, may or may not cap the cost of compliance, depending on whether the penalties are recoverable from customers. An obligated electricity supplier may prefer to comply with the RPS obligations, even if the cost of doing so exceeds the financial penalty, if compliance costs can be recovered from customers but financial penalties cannot. Such a situation is most likely to arise in still-regulated electricity markets not open to retail competition, in which case financial penalties are unlikely to cap the cost of the RPS. In markets with retail competition, on the other hand, financial penalties are more likely to be at least partially recoverable in rates, and therefore may be treated in the same fashion as SACPs by electricity suppliers, effectively capping the overall cost of the solar/DG set-aside.

[23] Under Illinois' RPS rules, competitive retail electric suppliers are required to meet at least 50% of their annual compliance obligation with ACP payments; these funds are transferred to the Illinois Power Authority, which uses the funds to procure RECs. The remaining 50% of each supplier's obligations can be met through any combination of ACPs, renewable energy, or REC purchases. ACP rates are updated annually based on the average cost of RECs procured through the IPA's solicitations for the state's IOUs. For compliance year 2009, ACP rates were equal to either $0.645 or $0.764 per MWh of retail sales, depending on utility service territory (or approximately $16 or $19 per MWh of renewable energy required). Illinois' solar set-aside requirement for competitive suppliers does not begin until 2015 (the set aside for the state's utilities begins in 2011-12), and whether there will be a unique solar ACP is unclear.

[24] The resurgent interest in CSP is also partly driven by the ability to incorporate thermal energy storage, and thereby provide greater energy market value than wind and other solar technologies.

[25] The apparent increase in compliance with Arizona's set-aside from 2007 to 2008 is largely an artifact of the state's transition from a solar set-aside to a DG set-aside, and the associated change in target levels and qualifying resources.

[26] The contribution of these solar resources to Nevada's set-aside is amplified by the application of credit multipliers for PV projects (2.4 for central-station PV and 2.45 for distributed PV).

[27] Some caution is warranted in using these data, as they do not include bilateral trade in SRECs or longer-term contracts, and because liquidity is limited in many states. With the exception of the NJ SREC data, it is unknown how many transactions underlie each data point shown in the figure.

[28] The SREC prices presented in Figure 9 do indicate that SREC prices have tended to be close to SACP levels, albeit somewhat lower. More generally, if SACPs are heavily used for compliance, then average SREC prices should tend approach the SACP.

[29] SREC pricing is also available for New Hampshire (see Figure 9), but compliance obligations for that state's solar set-aside do not begin until 2010, so retail rate impacts for 2009 are not included in Table 7.

[30] Specifically, retail rate impacts for Arizona are based on the DG set-aside expenditures reported by each applicable utility in its 2009 compliance filing, and retail rate impacts for New York are based on NYSERDA's 2009 budget for the RPS customer-sited tier. Neither of these states rely primarily on SRECs for compliance with its set-aside; thus, the cost impacts may not be directly comparable to the other states included in Table 7.

[31] This sum is in addition to California's policy goal of installing 3,000 MW of distributed PV by 2017. California's goal is not part of an RPS solar/DG set-aside, so is not included in Figure 10.

In: Solar Power, Renewable Portfolio Standards ... ISBN 978-1-62081-532-8
Editors: V. M. Evans and W. L. Reed

Chapter 2

SOLAR RENEWABLE ENERGY CERTIFICATE (SREC) MARKETS: STATUS AND TRENDS*

Lori Bird, Jenny Heeter and Claire Kreycik

LIST OF ACRONYMS

AC	alternating current
ACP	alternative compliance payment
BTU	British thermal unit
AC	alternating current
DC	direct current
EDC	electric distribution company
EGS	electric generation supplier
EY	energy year
IOU	investor-owned utility
kW	kilowatts
MW	megawatts
MWh	megawatt-hour
PJM	Pennsylvania New Jersey Maryland Interconnection
PSC	public service commission
PUC	public utility commission

* This is an edited, reformatted and augmented version of National Renewable Energy Laboratory, Publication No. NREL/TP-6A20-52868, dated November 2011.

PV	photovoltaic
RFP	request for proposal
RPS	renewable portfolio standard
SREC	solar renewable energy certificate
W	Watt

EXECUTIVE SUMMARY

Solar renewable energy certificates (SRECs) represent the generation attributes of solar energy systems and can be traded separately from commodity electricity. A SREC is created for each megawatt-hour of electricity generated from solar energy systems. SREC markets have emerged in nine U.S. states and Washington, D.C., as a method to meet compliance with solar carve out provisions of renewable energy portfolio standards (RPS). Solar carve out provisions set requirements for solar generation or capacity. This report examines the status and trends of SREC markets. Key issues and trends discussed in the report include:

- SREC markets are relatively young but expected to grow rapidly in coming years as state solar requirements ramp up. New Jersey has been the dominant market for SREC trading to date, but SREC markets in other states are expanding. Nationwide, solar carve outs in SREC markets are scheduled to grow from more than 520 MWAC in 2011 to nearly 7,300 MWAC in 2025.
- Policymakers establish key elements when designing SREC policies. Some states allow solar thermal to participate in SREC markets, but overall the SREC market is dominated by solar photovoltaics (PV). In terms of geographic eligibility, several states limit eligibility to in-state development, while a few allow SRECs from a broader geographic region.
- Solar alternative compliance payments (ACPs) generally set a ceiling on prices and are scheduled to decline over time, reflecting expectations of declining PV costs. In shortage situations, SRECs typically trade near the solar ACP.
- Lack of long-term contracts has been a barrier to project developers obtaining financing in some markets. This issue may continue to be a problem particularly in areas where rebates or other supplemental incentives expire and SRECs become the primary incentive stream for

financing systems. Several states have instituted provisions such as long-term contracting requirements, price floors, or standardized long-term contracts to try to overcome these challenges.

- Rate caps exist in some form in five SREC markets but have not been reached or are yet to be evaluated. The modest size of most solar carve outs suggests that it is not likely that rate caps will be reached in the near term.
- SREC markets are supporting a mix of PV system sizes, including residential, nonresidential, and utility-scale systems. In most markets, about two-thirds of registered projects are smaller than 10 kW. Projects larger than 250 kW dominate capacity installed in a few markets, including North Carolina, Massachusetts, and New Jersey. There has been a trend toward large-sized projects in recent years.
- Compliance has proved challenging in a number of SREC markets in recent years, but adequate supplies are expected going forward in many of these regions. A handful of states reported compliance shortfalls in 2009. The first compliance periods for a few markets occurred in 2010 or 2011. Recently, SREC spot prices have leveled or decreased in a number of markets as supplies have increased and compliance challenges have eased.
- SREC markets are expected to grow to more than 4,300 MW by 2020. In the near term, a sharp drop in SREC prices indicates that some markets may be oversupplied. An increase in installations has occurred in the last year due to a greater focus on U.S. markets, in large part due to slower growth in Europe, as well as a substantial drop in module prices. Capacity growth has been fueled in part by the approaching end of the Federal Treasury 1603 cash grant program, which will expire at the end of 2011. Looking forward, the largest markets to date are expected to slow over the next year or two but others are expected to continue to expand in the interim.

1. Introduction

Solar renewable energy certificate (SREC) markets have emerged in a number of U.S. states to encourage the adoption of solar energy systems. SRECs represent the generation attributes of solar energy systems and can be traded separately from commodity electricity. A SREC is created for each megawatt-hour (MWh) of electricity generated from solar energy systems.

SREC markets have emerged in some U.S. states that have renewable portfolio standards (RPS). These standards require utilities or load-serving entities (LSEs) to procure minimum amounts of renewable energy, including solar energy, to serve their electricity loads. Obligated entities demonstrate compliance with RPS solar energy requirements by retiring SRECs that they have acquired from the market.

RPS policies are widespread in the United States. By June 2011, 29 states, Washington, D.C., and Puerto Rico had adopted RPS policies, while another eight states had renewable portfolio goals. Of the jurisdictions with RPS policies in place, 17 had adopted specific targets, or "RPS carve outs," for solar energy or distributed energy resources to ensure that these sources would be part of the resource mix (Wiser et al. 2010). More than half of the 17 jurisdictions with solar carve outs in their RPS allow the use of SRECs to demonstrate compliance with solar targets. Other states have relied on financial incentives such as grants, rebates, performance-based incentives, feed-in tariffs, or standard offer contracts to achieve solar requirements or goals. In many cases, these financial incentives are used in conjunction with SRECs to help achieve compliance with solar carve outs.

SREC markets are relatively new. New Jersey was the first state to rely heavily on SRECs as a market mechanism to meet its solar carve out. New Jersey's first compliance year with its carve out was 2005, and initially, New Jersey established a rebate program to incentivize systems. Due to the high cost and constraints on the state budget, the state switched to a market-based SREC program in 2007 (Hart 2010). Since then, some states in the mid-Atlantic and surrounding regions have used SRECs to enable obligated entities to meet their solar carve outs, sometimes with other support schemes as well.

In several western states, SRECs are used to track compliance with solar carve outs, but there is no active trading market. In some cases, SRECs may be sold into voluntary markets in which consumers, businesses, and institutions purchase the renewable energy equivalent of their electricity needs. However, SRECs have played a limited role in these markets to date.

This paper examines experience in SREC markets in the United States. It describes how SREC markets function—key policy design provisions, eligible technologies, state and regional eligibility rules, solar alternative compliance payments (ACPs), measurement and verification methods, long-term contracting provisions, and rate caps. It also examines the trends of SREC markets—trading volumes, sourcing trends, trends in the size of solar photovoltaic (PV) systems driven by these markets, and trends in price and compliance. Throughout, the paper explores key issues and challenges facing

SREC markets and attempts by policymakers to address some of these market barriers. Data and information presented in this report are derived from SREC tracking systems, brokers and auctions, published reports, and information gleaned from market participants and interviews with state regulators responsible for SREC market implementation. The last section summarizes key findings.

2. SREC Market Overview and Design

2.1. Market Size and Scope

SREC Markets Have Expanded to 10 Jurisdictions

SREC markets have emerged or are in the early stages of implementation in 10 jurisdictions. While 17 jurisdictions have an RPS carve out requiring electricity providers to obtain a certain fraction of their generation portfolio from solar, distributed generation (DG), or customer-sited sources, not all of these states allow the use of SRECs for compliance. Figure 1 details the states that have an RPS solar carve out, and of those, which allow SREC trading.

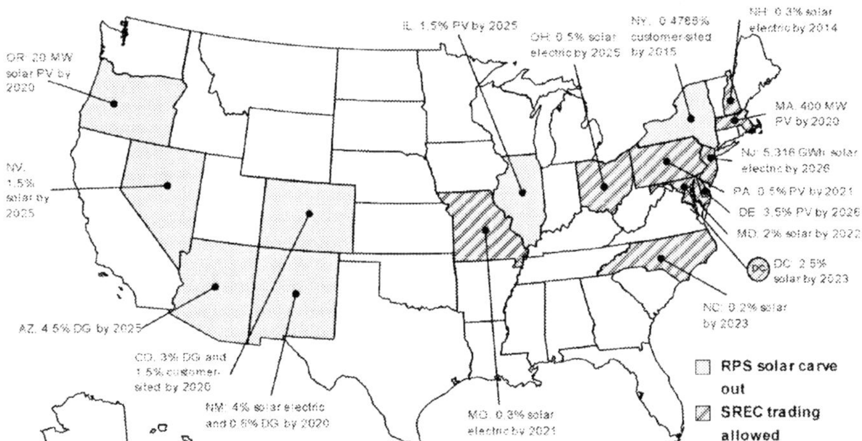

Figure 1. Map of states with SREC markets.

Trading in SRECs is relatively new. SREC markets began when state solar targets took effect in New Jersey and Pennsylvania in 2005 and in Delaware and Washington, D.C., in 2007 (see Table 1). SREC trading increased substantially in 2007 when New Jersey shifted from a rebate program to greater reliance on SREC markets to support solar projects. Since then, Maryland (2008), Ohio (2009), North Carolina (2010), New Hampshire (2010), and Massachusetts (2010) have begun implementation of their solar carve outs and associated SREC markets. Missouri's solar requirement took effect in 2011.

Table 1. State SREC Policy Overview

State	Initial compliance year	Target (% of Retail Sales)	Obligated Entities
DC	2007	2.5% by 2023	All EDCs and EGSs
DE	2007	3.5% by EY 2026	All EDCs Municipals/co-ops can set their own comparable RPS schedule by 2013
MA	2010	400 MW PV*	All EDCs and EGSs
MD	2008	2% by 2022	All EDCs and EGSs, municipals, co-ops
MO	2011	0.3% by 2021	All IOUs**
NC	2010	0.2% by 2018	All IOUs, municipals, co-ops
NH	2010	0.3% by 2014	All IOUs and retail suppliers, excluding municipals
NJ	2005	5,316 GWh by EY 2026	All EDCs and EGSs
OH	2009	0.5% by 2024	All EDCs and EGSs
PA	2005	0.5% by 2021	All EDCs and EGSs, voluntary participation by municipals, co-ops

Source: DSIRE 2011a.

Note: EY = energy year; EGS = electric generation supplier; EDC = electric distribution company; IOU = investor-owned utility.

*Massachusetts's target is adjusted based on market conditions during previous years. In 2010, the compliance obligation was 34,164 MWh, the generation equivalent of 30 MW of capacity operating at a 13% capacity factor. This is equivalent to 0.07% of retail sales.

** Empire District Electric, one of Missouri's four IOUs, is exempt from the state's solar energy standard because it installed renewable energy capacity equal to 15% of the utility's total owned fossil-fired generating capacity by January 2009 (DSIRE 2011a).

SREC targets vary considerably depending on the goals of each jurisdiction. The ultimate target for the percent of retail sales that must be met by solar energy varies from 0.2% by 2018 in North Carolina to a high of 3.5% by 2026 in Delaware (see Table 1 and Figure 1). If converted to a percentage of retail sales, New Jersey's target is even higher. The stringency of the solar target, as well as other policy features, determines the extent to which new solar development is encouraged.

In terms of total megawatts required under SREC policies, New Jersey is by far the largest market, requiring approximately 319 $_{MWAC}$ of solar capacity in 2011 and projected to have more than 4,000 $_{MWAC}$ by 2025 (see Table 2 and Figure 2). The next largest near-term markets are Massachusetts, Pennsylvania, and Ohio. The Maryland market, while relatively modest in 2011, has aggressive targets that would require it to increase rapidly, making it the fourth largest market by 2015 and second largest market after New Jersey in 2020.

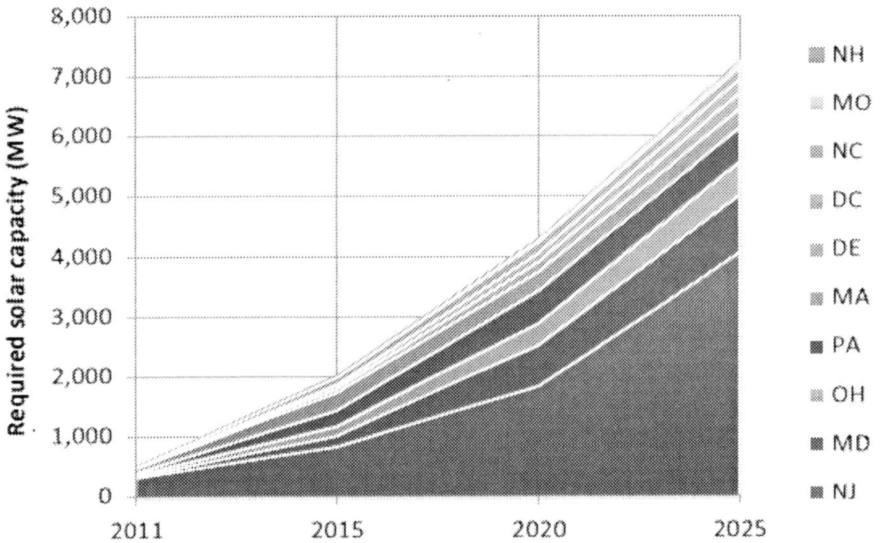

Source: Barbose 2011, with updates.

Figure 2. Future capacity required in SREC markets (2011–2025).

SREC markets began in restructured[1] states in the mid-Atlantic and Northeast but have since expanded into states with regulated electricity markets. In restructured states, where organized electricity markets are in place, multiple LSEs may need to procure SRECs to meet RPS requirements.

The responsibility of satisfying the RPS requirements is placed on both electricity distribution companies (EDCs) that supply default service to customers and alternative suppliers, known as electricity generation suppliers (EGSs). In these markets, SRECs have been a useful tool. They enable trading among parties and they enable EGSs to procure SRECs from the spot market on a short-term basis or bi-laterally on a long-term basis. This flexibility of acquiring SRECs is important to both EDCs and EGSs because of uncertainty in future loads due to retail choice.

Table 2. Capacity Required in SREC Markets (in MW_{AC})

State	2011	2015	2020	2025
DC	>3	63	146	240
DE	10	69	162	261
MA	69	299	312	312
MD	22	179	693	958
MO	11	28	59	93
NC	18	121	181	190
NH	6	22	23	24
NJ	319	829	1,825	4,053
OH	31	155	363	553
PA	33	264	546	567
TOTAL	>522	2,029	4,310	7,251

Source: Barbose 2011, with updates.

Note: Most states have annual targets based on a percentage of retail electric sales in the compliance year. These estimates use the megawatt-hour target and convert it to megawatts using default capacity factors in NREL's PVWatts model, which determines the energy production of grid-connected PV systems. The capacity factors used assume that systems are south-facing and that the tilt is equal to the latitude of the state. The Massachusetts ultimate target, while not based on a percent of retail sales, requires a cumulative installment of 400 $_{MWDC}$ of solar capacity. These figures assume that Massachusetts's solar requirement are not adjusted for under- or over-supply.

Note: In Washington, D.C., the 2011 compliance obligation is uncertain because contracts entered into before July 12, 2011, are exempt from the District's increased solar requirement. This analysis assumes that existing contracts will expire before 2015 (Council of the District of Columbia 2011).

Several states with traditionally regulated electric utilities, including Missouri and North Carolina, allow the use of SRECs to achieve compliance with RPS targets. This enables utilities to procure SRECs from a broader geographic region than the state or balancing area. In a few states, municipal utilities and electric cooperatives are also obligated (or encouraged) to comply with RPS standards (see Table 1), in some cases with different targets.[2]

States measure RPS and solar set-aside compliance by looking at SRECs retired by the obligated entities. SRECs are retired when the owner designates that the SRECs will be used for compliance with a particular state RPS and transfers the SRECs to a retirement subaccount in the tracking system used to verify compliance. Regulators consider a 12-month span referred to as an energy year (EY) or compliance year. For example, the 2010 compliance period may consider SRECs retired between January 1, 2010, and December 31, 2010, or June 1, 2009, to May 31, 2010. Delaware, New Jersey, and Pennsylvania use a June 1 to May 31 period, while D.C., Maryland, Massachusetts, New Hampshire, North Carolina, and Ohio use a January 1 to December 31 period (SRECTrade 2011a).

2.2. System Eligibility: PV and Solar Thermal

SREC Market is Dominated by PV, Although Solar Thermal Is Eligible in Some States

A key policy design feature of solar carve outs that impacts SREC markets is technology eligibility. PV is an eligible—and most commonly used—technology in meeting solar carve out requirements in all SREC markets. Eligibility of solar thermal technologies is more varied.

About half of the state SREC markets allow solar thermal systems to participate (Table 3). Maryland, North Carolina, Ohio, and Washington, D.C., allow solar thermal technologies, including solar water heating, space heating and cooling systems, and other applications to participate in the SREC market. Solar thermal systems in Delaware, Missouri, and Ohio must generate electricity in order to be eligible.[3] In Pennsylvania, solar thermal that generates electricity is eligible for the primary resource targets of Pennsylvania's RPS, but it is not eligible for the state's solar carve out.

Despite its eligibility in a number of states, solar thermal has played a limited role in SREC markets to date. Based on interviews with state regulators, few solar thermal projects have been used in SREC markets in Missouri, Ohio, and Washington, D.C. As of October 2011, 287 solar thermal projects have been registered in PJM-GATS; the majority of systems (75%) are located in Virginia and North Carolina (PJM-GATS 2011a). For example, in North Carolina, the public utilities commission had approved 50 solar thermal projects, primarily solar hot water systems, compared to 205 PV projects (NCUC 2011a).

Table 3. Technologies Eligible to Participate in SREC Markets

State	Solar PV	Solar Thermal
DC	Yes	Yes
DE	Yes	Yes, but must generate electricity
MA	Yes	No
MD	Yes	Solar water heating, but must be installed on or after June 1, 2011
MO	Yes	Yes, but must generate electricity
NC	Yes	Yes*
NH	Yes	No**
NJ	Yes	No
OH	Yes	Yes, but must generate electricity
PA	Yes	No***

Source: SRECTrade 2011a

* Eligible solar thermal resources include solar hot water, solar absorption cooling, solar dehumidification, solar thermally driven refrigeration, and solar industrial process heat.

** Solar water heating that displaces electricity is eligible for Class I of New Hampshire's RPS (Epsen 2011).

*** Solar thermal is eligible for Tier 1 of Pennsylvania's RPS.

States have expanded eligibility for use of solar thermal over time, indicating that there may be more development of solar thermal in the future. In December 2010, under the "Solar Collector Certification Temporary Amendment Act of 2010," the Washington, D.C., SREC market was expanded to include non-residential solar thermal systems, whereas previously only residential solar thermal systems were eligible. In May 2011, Maryland passed legislation enabling solar water heating systems to be eligible in the Maryland SREC market. Systems installed on or after June 1, 2011, are eligible.

2.3. State and Regional Eligibility Rules Define Markets

Geographic Eligibility Requirements Influence the Location of Development

Another key policy design feature of solar carve outs that impacts SREC markets is limitations on the geographic eligibility of systems. Table 4 summarizes state geographic eligibility rules. Some jurisdictions limit the eligibility of solar to in-state projects through interconnection or other requirements (Delaware, Massachusetts, and New Jersey), while others allow a portion of the projects to come from out of state (North Carolina and Ohio). Other states allow SRECs sourced from projects in a specified region, such as the electric balancing area (Pennsylvania), or allow regional resources if in-state resources are insufficient (Maryland, though a binding in-state requirement goes into effect after 2011) (DSIRE 2011b). Washington, D.C., previously allowed regional resources to be used if local resources were insufficient; however, legislation enacted in August 2011 specifies that solar generating systems must be located within D.C. or in locations served by a distribution feeder serving D.C. Missouri is unique in that 100% of its solar standard can be met with SRECs from anywhere in the United States. North Carolina allows SRECs from anywhere in the United States to meet up to 25% of its solar carve out.

Table 4. Geographic Eligibility Requirements in SREC Markets

State	Geographic Eligibility
DC	In-district or in locations served by a distribution feeder serving the District
DE	Customer-sited solar must be located in state; non-customer sited solar can be located with PJM, or show import capabilities into PJM
MA	In state
MD	Solar resources must be connected with the distribution grid serving Maryland. On or before December 31, 2011, solar resources not connected to the Maryland grid are eligible if electricity suppliers were unable to contract for in-state SRECs
MO	Out-of-state eligible (anywhere)
NC	25% from out-of-state eligible (anywhere)
NH	Within New England or New York
NJ	In state
OH	50% of SRECs must be generated by in-state resources; bordering states (DC, IN, KY, MD, MI, PA, WV) and areas able to prove delivery to Ohio are eligible for 50%
PA	Out-of-state eligible (within PJM)

Source: DSIRE 2011a, Epsen 2011.

Because only a few states allow SRECs to come from a broad geographic region, the national market for SRECs is limited. Generally, for nationally sourced SRECs to be eligible to qualify for RPS compliance, systems must be registered in renewable energy certificate (REC) tracking systems. For example, California and other western states have supplied SRECs registered in the Western Renewable Energy Generation Information System (WREGIS) tracking system to North Carolina and Missouri to help meet their solar requirements. There are currently nine REC tracking systems in operation in the United States.

In California, the state with the most grid-connected PV capacity, system owners retain the RECs and can resell them into SREC markets, even if they receive financial incentives under the California Solar Initiative.[4] The California Public Utilities Commission has issued a proposed decision that would allow system owners who install PV systems under the California Solar Initiative program to sell their SRECs to California utilities to meet the portion of the new 33% by 2020 RPS that can be met with RECs (CPUC 2011).

Some states have faced challenges regarding RPS geographic eligibility rules on the basis of the Dormant Commerce Clause of the U.S. Constitution, which precludes states from discriminating against or impeding interstate commerce. For example, Massachusetts faced a Commerce Clause challenge to its solar requirement, resulting in a partial settlement whereby the in-state requirement remains in place with modifications to its solar price cap, or solar alternative compliance payment (ACP), mechanism among other changes (Elefant and Holt 2011). However, in-state solar or DG requirements are untested by the courts to date.

The design of geographic restrictions may have bearing on the constitutionality. According to a recent legal analysis, solar or DG carve outs that are limited to in-state resources are "facially discriminatory," or discriminate on their face (Elefant and Holt 2011). They must be justified by demonstrating that the state lacks alternatives to achieve legitimate policy goals (Elefant and Holt 2011). However, eligibility based on functional requirements such as interconnection, deliverability, or displacement of power, which can be functionally equivalent to in-state requirements, might not be viewed as discriminatory, and therefore, has lower risk of Commerce Clause challenges. States may be able to justify such policies because the size of requirements would place a minimal burden on commerce and offer legitimate benefits to the state, such as reliability and diversity of supplies that may not be readily achieved through other means (Elefant and Holt 2011).

2.4. Solar Alternative Compliance Payment Levels

Solar ACPs Generally Set a Ceiling on Prices and Are Scheduled to Decline in Most Jurisdictions

The price of SRECs is generally limited by the solar ACP established by each state with a SREC market. If an entity obligated to meet the RPS cannot procure enough SRECs, it can pay the solar ACP for each megawatt-hour that it has not procured in the market. For this reason, the solar ACP essentially sets a ceiling on SREC prices, as obligated entities would not pay more than the level of the ACP to acquire SRECs for compliance. However, if SREC purchases are recoverable in rates and ACP payments are not, such as in Ohio and Missouri, it is possible that SREC payments could exceed the solar ACP. Payment of the ACP is generally the mechanism of last resort to achieve compliance with the RPS.

In most SREC jurisdictions, a solar ACP has been established either through legislation or by the state public utility commission (PUC). The solar ACP differs from the ACP established for RECs used to meet the main RPS resource targets. Solar ACPs are set at levels higher than main RPS ACPs to reflect the higher cost of solar energy and to enable adequate support for solar projects. In situations where obligated entities are having difficulties in meeting solar RPS targets, SREC prices rise to levels near the solar ACP, and in oversupply situations, the price of SRECs can fall well below the solar ACP.

Differences in solar ACP levels have affected SREC trading prices. For example, prices have been highest in New Jersey because New Jersey's solar ACP of nearly $700/MWh is significantly higher than solar ACPs in other states, which are in the $400–$500/MWh range. In Pennsylvania, solar ACPs are set at approximately twice the price paid for SRECs. In Missouri, there is a provision for non-compliance equal to twice the average market value of SRECs utilized for compliance during the period; non-compliance payments are not recoverable from ratepayers. North Carolina is unique in that there is no specific penalty or formula for assessing alternative compliance; it is up to the North Carolina PUC to enforce compliance. While Illinois has a significant solar carve out in its RPS, which takes effect in 2013, it is unclear if there will be a separate solar ACP (Wiser et al. 2010), which would be necessary to encourage a distinct SREC market. If there is not, SRECs would not trade above the price of the main RPS ACP.

Most states have established long-term schedules for solar ACPs that generally decline over time given the expectation that the price of solar will fall in coming years, consistent with recent trends (see Table 5). The availability of long-term solar ACP schedules helps provide some certainty to the market, enabling developers and financiers to price forward contracts. Solar ACPs decline over time in New Jersey, Ohio, and Maryland, and in Massachusetts, the state is authorized to reduce the rate but not by more than 10% in a compliance year (Table 5). Massachusetts recently decreased the solar ACP from $600/MWh to $550/MWh and proposed future levels for the next 10 years.

Table 5. State Solar Alternative Compliance Payments

State	Solar Alternative Compliance Payment
DC	$500/MWh through 2016, $350/MWh in 2017, $300/MWh in 2018, $200/MWh in 2019 and 2020, $150/MWh in 2021 and 2022, and $50/MWh in 2023 and thereafter
DE	$400/MWh for LSEs first year of use, $450/MWh in second year, $500/MWh in third and subsequent years
MA	$600/MWh in 2010; $550/MWh in 2011, 2012, and 2013; and proposed to decrease 5% each year through 2021
MD	$450/MWh for solar shortfalls in 2008, $400/MWh in 2009 through 2014, $350/MWh in 2015 and 2016, $200/MWh in 2017 and 2018, and continuing to decline by $50/MWh bi-annually until it reaches $50/MWh in 2023 and beyond
MO	Penalty of at least twice the average market value of SRECs for the calendar year
NC	No specified penalty but the North Carolina PUC is authorized to enforce compliance
NH	$163/MWh in 2011, adjusts annually based on Consumer Price Index
NJ	$693/MWh in EY 2010; declines by $16/MWh–$18/MWh annually through 2015
OH	$450/MWh in 2009, $400/MWh in 2010 and 2011; declines by $50/MWh every two years to meet a minimum of $50/MWh in 2024[a]
PA	Roughly twice the price paid for PV by electricity suppliers in the state[b]

[a] The Ohio PUC has the ability to increase the solar ACP if it determines entities are defaulting to the ACP rather than pursuing the necessary resources.

[b] The calculation also takes into account solar rebates, but this represents a small portion of the solar ACP. Source: DSIRE 2011a.

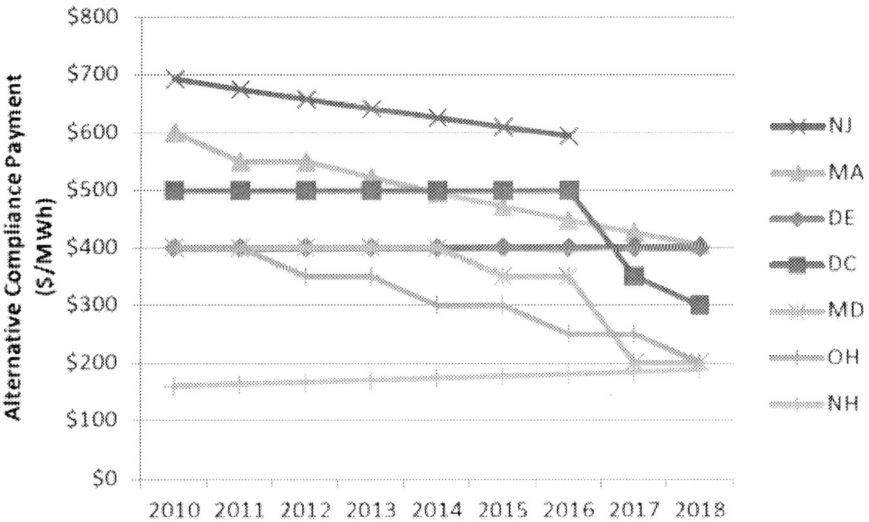

Source: DSIRE 2011a.

Note: Massachusetts solar ACP levels for 2010 and 2011 are fixed; levels for 2012 and beyond have been proposed by the Department of Energy Resources but are not yet finalized.

Figure 3. Solar alternative compliance payment levels over time.

2.5. Measurement and Verification Methods for SRECs

Methods of Verifying the Generation of SRECs Vary by System Size

Because SRECs can sell at high prices, an important issue is verifying the output from solar generators to ensure that the SRECs that are created match actual generation. Requirements for SREC measurement and verification vary to some degree across each state market. However, in general, most states allow small systems of less than about 10 kW to estimate annual generation based on models or engineering calculations. For example, models, such as PVWatts, are used to estimate the output of a facility based on system characteristics and average insolation for a particular location (Table 6). Typically for larger systems, a revenue grade meter is required to verify actual generation. This information is then self-reported to the REC tracking systems in most instances. Massachusetts has delegated the authority to independent monitors to report production data for participating SREC systems greater than 10 kW.

Table 6. State Requirements for Verifying Solar Output to Sell SRECs

State	Facility Reporting Provisions
DC	Facilities <10 kW can use estimated generation based on PVWatts calculation; otherwise, facilities can self-report via a revenue grade meter. Energy output of solar thermal systems >10,000 kWh per year must be rated and certified, and energy output must be determined by an onsite energy meter. Solar thermal systems ≤10,000 kWh per year may use an onsite energy meter or the annual performance measure determined by SRCC OG-300.
DE	Self-reporting via a revenue grade meter is used.
MA	All PV projects must self-report via a revenue grade meter to an independent verifier, the Massachusetts Clean Energy Technology Center. Projects <10 kW can report manually, while projects >10kW must report production using a data acquisition system (DAS). Currently there are 20 such DAS providers listed on the Massachusetts RPS website. The Massachusetts Clean Energy Technology Center maintains the production tracking system, verifies production data, and submits reports quarterly to NEPOOL-GIS.
MD	Facilities <10 kW may use estimated generation based on the PVWatts calculator. All other facilities must report monthly meter readings. Solar hot water systems must be certified and metered.
MO	Not yet determined.
NC	Projects installed "behind the meter" will be able to self-report their energy production data into the NC-RETS system to receive SRECs, while utility-metered projects will require a designated "Qualified Reporting Entity" with a separate NC-RETS account to upload generation data on their behalf. Systems >10 kW must be metered by the customer.
NH	All projects installed on the customer side of the meter, regardless of size, require an independent monitor for production verification.
NJ	Facilities <10 kW may use estimated generation based on the PVWatts calculator. All other facilities must report monthly meter readings.
OH	Self-reporting via revenue grade meter is used. If the system is ≤6 kW, generation can be reported from an inverter meter.
PA	Facilities >15 kW must report monthly meter readings. Facilities <15 kW are eligible to use estimated generation based on PVWatts calculation, unless they receive funding from the PA Sunshine Fund.

Sources: SRECTrade 2011a; Massachusetts DOER 2011a; Epsen 2011.

States allowing solar thermal systems use a conversion factor to issue SRECs to solar thermal projects. Converting non-electric solar thermal output into a SREC equivalent is done using a conversion factor of 3,412,000 British thermal units (BTUs) to 1 SREC. This standard rate is used by states but only specified in implementation rules by North Carolina. In Washington, D.C., smaller systems use the kilowatt-hour (kWh) factor supplied by the Solar Rating and Certification Corporation. Larger systems in Washington, D.C., must be metered; the BTU output is then converted to kilowatt-hours. Maryland specifies that BTUs created by solar thermal systems be converted to kilowatt-hours.

2.6. Long-Term Contracting and Financing Provisions

Lack of Long-Term Contracts Have Impeded Development in Some Cases: Some States Have Developed Mechanisms to Help Facilitate Financing

In some instances, reliance on spot market SREC prices has led to challenges in achieving compliance with solar carve outs, particularly in restructured electricity markets. Reliance on short-term spot markets for SRECs can pose problems for solar developers in obtaining the financing necessary to bring projects to fruition. Financiers (both debt and equity lenders) generally do not take SREC revenues into account unless a contract with a creditworthy purchaser is in place (Holt et al. 2011). This issue can be particularly problematic for large-scale systems that require debt or equity financing. At the same time, many electricity suppliers, especially in restructured electricity markets, resist signing contracts longer than two to three years. This may be because future loads, and therefore future solar compliance obligations, are unknown. Additionally, suppliers in regulated states may perceive some risk regarding cost recovery mechanisms if the contract extends beyond the current rate plan. As a result, some solar developers have had difficulty obtaining financing for their projects due to a lack of opportunities to enter long-term SREC contracts.

To address challenges in obtaining financing, some state policymakers have implemented provisions that reduce investor risk. As noted earlier, long-term solar ACP schedules can help reduce investor risk by providing knowledge of future price caps. In SREC markets, other provisions include: (1) requiring or encouraging long-term contracts or (2) establishing payment floors to set a minimum price for SRECs, which guarantee a minimum return. Also, some jurisdictions have implemented price responsive demand mechanisms through which future solar targets are reduced in the case of shortages or increased in the case of oversupplies. Also, many utilities in regulated electric markets offer long-term contracts for the SRECs and output of solar energy systems to facilitate development or elect to own solar projects.[5] In some jurisdictions, rebates or other types of incentives have been offered to reduce the risk and need to rely solely on SREC revenues, but we do not cover these programs in detail below.

Long-Term Contracting Requirements

State policymakers and regulators in New Jersey, Pennsylvania, and to a lesser extent Maryland encourage the use of long-term contracting to enable solar projects to obtain financing.

New Jersey has created a formal SREC-based financing program that awards long-term contracts through requests for proposals (RFPs) to help winning residential and commercial developers secure financing for their solar projects. The contracts resulting from the RFPs are standardized and non-negotiable, and developers are encouraged to bid SREC prices that would provide sufficient project returns. These programs represent only a fraction of the overall market, which called for about 320 $_{MWAC}$ of installed capacity in 2011 alone. The Jersey Central Power & Light, Atlantic City Energy, and Rockland Electric Company SRECbased financing programs were designed to support a total of 64 MW (name plate) from 2009 through 2011 (NERA Economic Consulting (2011b). The PSEG Solar Loan program had the goal of supporting 81 MW in the first two phases (NJ BPU 2011 d) (see Appendix for additional detail). As a result of these programs and perhaps other market conditions, longterm contracts have historically been more widely available in New Jersey than in other markets, making it easier for developers to operate and obtain financing. However, the availability of long-term contracts outside of the utility programs has declined recently with the drop-in SREC prices and signs of near-term oversupply in the market, as one might expect (SEIA/GTM 2011b; SEIA/GTM 2011c).

In Pennsylvania, the PUC issued a policy statement encouraging EDCs to conduct competitive solicitations for SRECs and procure SRECs from small-scale projects (<200 kW) through RFPs and bilateral contracts. It specifies that the price negotiated for SRECs from small-scale projects should not exceed the average winning bid price in the EDC's most recent RFPs for large-scale solar projects (Pennsylvania PUC 2011). In response, Pennsylvania EDCs have started to hold competitive solicitations to procure a portion of their SREC obligations over a longer delivery period (e.g., 7–10 years). The utilities are seeking bids in tranches (i.e., 100, 250, or 500 SRECs per year from an individual supplier). One of the utilities, PPL Electric Utilities, is issuing RFPs for individual projects as well as aggregations of projects <15 kW in size.

These programs have not been large to date, however. The Met-Ed and Penelec solicitiations in 2010 were for 25,000 MWh while Penn Power and PPL Electric solicitations in 2011 were for a total of 46,000 MWh (see Table 10 and Appendix for additional details). The state is well ahead of its compliance targets, but this has been largely attributed to the availability of rebates that created an attractive incentive for projects (SEIA/GTM 2011b).

Maryland has a loose requirement aimed at helping buyers and sellers find each other and negotiate bilateral contracts. The Maryland Public Service Commission (PSC) requires solar generators to first offer their SRECs for sale on the PSC's website, if they want to sell directly to Maryland electricity suppliers requests for proposals. If the electricity suppliers enter into contracts directly with generators, the contracts must be for 15 years. Because long-term contracts are not required for purchases from brokers, most of the transactions in the market are broker-facilitated spot transactions (Mosier 2011). Therefore, this non-binding contracting requirement has had little effect in encouraging long-term contracts to date. Developers have indicated that the lack of long-term contracts makes it more difficult to secure financing for projects in Maryland because banks will not consider SREC revenues without a contract (SEIA/GTM 2011b). Therefore, developers have to take on more of this financial risk.

In Delaware, a long-term contracting program has been proposed. Delmarva Power applied to the Delaware PSC for the approval of a pilot SREC procurement program on September 16, 2011. Under the program, solar generators of different sizes could receive fixed SREC prices for 20 years. For example, small systems under 50 kW would receive $260/MWh for 10 years, followed by payments of $50/MWh for another 10 years. Larger systems of 250 kW to 2 MW would receive a bid price, if they submit winning bids. The program is slated to support a total of about 11,500 MWh, divided among four system size categories (Moore 2011).

Long-Term Price Floors

Long-term price floors have been implemented in Massachusetts and New Jersey by the distribution utility PSE&G (see Appendix for additional detail on these mechanisms). Theoretically, long-term price floors serve the same purpose as long-term contracts. If a long-term price floor is structured in a way that guarantees a minimum value for SRECs in future years, investors may view it as they would a binding SREC contract.

In Massachusetts, the Solar Credit Clearinghouse was established in 2010 with the goal of establishing a minimum price to encourage development. The program enables any unsold SRECs in a compliance period to be auctioned at a minimum price of approximately $300/MWh.

The useful life of SRECs is extended once they are submitted into an auction account, and the lifetime can be extended further if the SRECs do not sell at auction initially. So far, the program is untested because no auctions have been held due to shortages in the market. Despite its complexity and initial reservations about the program, investors are becoming more comfortable with this market design (SEIA/GTM 2011b). However, the program structure does not guarantee the sale of SRECs if the final auction does not clear. Also, it does not guarantee cash flows in a particular year. For example, if a SREC is not cleared during an auction and its shelf-life is extended and it is sold in a later year, the inability to receive payments in the first year can impact the return on a project.

In New Jersey, the PSE&G loan program has an element of a price floor to provide confidence to consumers regarding the repayment of the loan. Under the program, PSE&G offers loans that cover 40%–60% of the cost of a PV system. Customers may repay the loan principal and interest through cash payments or by signing their SRECs over to PSE&G.

Throughout the 15-year loan term, PSE&G credits customers for each SREC generated based on market prices but not below the established price floor, which is initially $420/MWh for residential systems. While spot prices have historically been much higher than these levels, this situation may be changing. Forward prices for 2012 recently fell well below the $420/MWh price floor, which may mean that the utility will have to recover the price differential from ratepayers. However, this remains to be seen.

Price Responsive Demand

The Massachusetts program also incorporates mechanisms to adjust the solar targets based on compliance with targets. These provisions can help regulate the supply and demand balance and reduce risks of an oversupplied market. For example, under the Massachusetts Solar Credit Clearinghouse program, the term that SRECs can be deposited in the auction is adjusted based on market conditions. Initially projects can opt to deposit SRECs in the auction account for up to 10 years of the life of the system, but that can be reduced if the market is oversupplied.

Additionally, the compliance targets in Massachusetts are increased by the amount of SRECs deposited in the auction in a given compliance year to temper situations of oversupply (see Appendix for additional details).

Utility Standard Offer or Long-Term Contracts

In traditionally regulated or hybrid electricity markets, such as North Carolina, Missouri, and Ohio, utilities offer fixed long-term contracts to solar generators to facilitate in-state solar development. Standard offer contracts can help the economics of small to mid-sized projects in particular. They provide certainty regarding the price that consumers can obtain for their SRECs, although they may not significantly influence the ability of retail customers to obtain loans for their projects. In general, payment terms have ranged from $30/MWh to $300/MWh for terms ranging from 5 to 20 years. Most of these programs have been offered on a trial basis, with limits on the enrollment period (one or two years) or limits on the overall capacity that can participate in the program (see text box for examples).

Overall, the New Jersey long-term contracting provisions have been effective in driving new renewable energy development and enabling projects to obtain financing. However, falling SREC prices may mean that ratepayers will need to pay the difference between the long-term contracts offered by the distribution utilities and market prices. The price floor mechanism in Massachusetts has yet to be fully tested as the auction mechanism has not yet been used, but developers and investors may be gaining comfort with this model, according to industry analysts (SEIA/GTM 2011b). Fixed long-term contracts will continue to be important to drive in-state solar energy development in other markets; however, as many of the existing programs have limited funds or are available for a limited duration, it is unclear if they will be extended in future years.

Sample Utility Long-Term SREC Purchasing Programs to Encourage Small to Mid-Sized Systems

As a method of procuring SRECs, several utilities in Missouri, North Carolina, and Ohio have developed SREC long-term purchasing programs for retail customers. These programs are either structured as standard offer contracts for SRECs, or upfront incentives, which are awarded in exchange for the right to keep the SRECs for a pre-defined number of years.

Ameren Missouri was the first utility to develop a standard offer contract program for solar in Missouri. Customers installing systems with a nameplate capacity of less than 10 kW may sell SRECs for $100/MWh on a 10-year upfront basis, based on PVWatts estimates of output.

Systems between 10 kW up to 100 kW in size may sell SRECs for $100/MWh on a 5-year basis based on actual generation from their metered output (Ameren 2011b). The program, which is separate from the solar rebate required by law, was funded with $2 million for 2011, and all funding was committed as of mid-May 2011 with contracts completed or pending for 2,269 SRECs (Ameren 2011a). Ameren may continue the program in 2012 and 2013, but it has not been submitted to the PUC (Ameren 2011a; Ameren 2011b).

Duke Energy Carolinas launched a standard offer purchase program for RECs including SRECs in North Carolina in 2008. Under the program, the utility enters into contracts ranging from 5 years to a maximum term of 15 years. Contracts must be for at least 35 MWh/year of SRECs (approximately the output of a 25-kW system). Solar systems that generate more than 250 MWh/year are ineligible and can submit an unsolicited bid proposal to the utility instead. In 2010, customers were paid $30/MWh for SRECs under the standard offer program (this represents the price for the SRECs only, not the underlying electricity). SREC prices escalate annually for each standing contract. Prices will ultimately increase to $42/MWh in 2025 (Duke Energy Carolinas 2011).

In Ohio, **Duke Energy Ohio** offers a residential PV REC purchase program that was approved in July 2010. The program is available through December 31, 2012, to Duke's residential customers, including both customers purchasing their electric generation service from Duke and those purchasing from other retail suppliers. The REC purchase agreements have a term of 15 years. Customers received $300/MWh for SRECs procured in 2010, and payment levels will be adjusted in subsequent years based on market prices (Duke Energy Ohio 2011a).

Progress Energy Carolinas (Progress) has a similar standard purchase offer for North and South Carolina non-residential customers called the SunSense Commercial PV Program. Progress pays $180/MWh for both the electricity and SRECs generated by the PV system for a period of at least 20 years. To be eligible, the PV systems must have a nameplate capacity between 11 $_{\text{kWDC}}$ and 500 $_{\text{kWDC}}$. Annual program participation is limited to 5 MW and net metering is not offered. Systems owned and operated by a third-party on a Progress Energy customer's property are also eligible. For residential systems sized 2 $_{\text{kWAC}}$ to 10 $_{\text{kWAC}}$ the utility provides a $1/watt$_{\text{AC}}$ upfront incentive and a monthly bill credit based on the AC rating of $4.5/kW (Progress Energy Carolinas 2011b).

First Energy ran a residential REC purchasing program in Ohio in the service territories of Ohio Edison, The Illuminating Company, and Toledo Edison's between 2009 and May 31, 2011. Under the program, First Energy offered 15-year contracts for residential SRECs. The price paid was determined through a biannual RFP process. The PUCO and FirstEnergy determined that if the utility was unable to procure SRECs through competitive solicitations, an alternative purchase price would be offered to residential customers. The price offered for SRECs from residential systems was \$360/MWh in 2009 and \$320/MWh in 2010 and 2011, decreasing by \$40/MWh every two years over the lifetime of the contract (DSIRE 2011c).

AEP Ohio began offering an incentive program for both residential and non-residential customers called the Renewable Energy Technology Program Rider in July 2011. The two-year program is funded at \$1 million annually for solar. The current solar incentive is \$1.50 per Watt, and this payment is provided to customers in exchange for the SRECs generated by the project for a 15-year period (AEP Ohio 2011).

2.7. Rate Caps

Rate Caps Have Yet to be Reached Although Policy Implementation is in Early Stages

Going forward, the need for obligated entities to meet solar carve outs can be influenced by the presence of electricity rate impact caps. These caps were established by legislators to limit the ratepayer impacts of RPS policies or, in some cases, the solar carve out in particular.

Rate caps exist in some form in five SREC markets (Maryland, Delaware, Ohio, Missouri, and North Carolina) but have not been reached or are yet to be evaluated (Table 7). Delaware and Maryland have rate caps on the cost of the solar set-aside requirement, while other states have caps on the entire RPS.

In Delaware, the cost of compliance includes costs associated with any ratepayer-funded rebate program, SREC purchases, and solar ACPs. If these costs exceed 1% of the total retail cost of electricity, then the PSC may freeze the solar requirement; Delaware also has a 3% cap on its RPS (14 DE Reg 1241, 3.2.16). This provision went into effect in July 2010, so the impact cannot be assessed until after utility compliance reports are filed in September 2011 (Stewart 2011).

Table 7. Rate Cap on RPS or Solar Set-Aside

Rate Cap on RPS	Rate Cap on Solar Set-Aside	No Rate Cap
OH: 3% increase in generation costs	DE: 1% increase in retail rates (combination of solar set-aside, rebates, and solar ACPs together)*	
NC: Caps on annual cost per account for incremental RPS costs; varies by customer class and year	MD: 1% increase in retail rates (solar set-aside only); allows for 1-year delay in meeting solar RPS	DC, PA, MA, NH, NJ**
MO: 1% increase in retail rates		

Source: Wiser et al. 2010; DSIRE 2011a; N.J.A.C 14:8-2.2 and 2.3, revised effective March 30, 2011. * Delaware also has a 3% cap on its RPS.

**New Jersey previously had a rate cap of 2%, but the provision was removed as of March 2011.

In Maryland, the increase is based on the actual or projected dollar-for-dollar cost to purchase SRECs. If that cost is 1% or more of a supplier's total annual electricity sales revenues, then compliance can be delayed by one year (DSIRE 2011a). No requests have yet been made to delay compliance in Maryland (PSC MD 2011).

Ohio, Missouri, and North Carolina have rate caps on the entire RPS policy. In Ohio, the increase is based on "reasonably expected" rate increases of no more than 3%. If rates are expected to increase by 3% or more above what they would have otherwise been, the cap goes into effect (DSIRE 2011a).

In Missouri, utilities may be excused from their obligation if the cost of complying with the RPS increases retail electricity rates by more than 1% in any year. If this occurs, the requirement will be adjusted in that year so that the rate cap is not violated (DSIRE 2011a).

In their RPS planning dockets, no utilities in Missouri indicated potential shortages in SREC supply, likely due to the fact that they can obtain SRECs from out-of-state markets. One utility, Ameren, noted that it was able to procure SRECs at an average price of about $8/MWh. Ameren's total projected compliance cost in 2011 was $5.3 million, which represents 0.01% of Ameren's total revenue requirement (Ameren 2011b).

Similarly, Kansas City Power & Light and KCP&L Greater Missouri Operations Company estimated that the average increase in annual revenue requirement would be 0.14% on a 3-year average and 0.04% on a 10-year average (KCP&L-GMO 2011), which is far less than the 1% rate cap.

North Carolina's RPS legislation takes a different approach to capping costs. Utilities can recover up to a maximum dollar amount of "incremental costs" annually from each customer class (residential, commercial, and industrial). In 2011, for example, utilities were allowed to recover a maximum of $10 per residential account, $50 per commercial account, and $500 per industrial account (N.C. Gen. Stat. § 62-133.8(h)(4)).

In the future, it is possible that rate caps may be reached; whether they will be reached will depend on many factors, including future SREC costs and overall retail rates. Generally, the likelihood of reaching a cap is greater in states that cap the impact of the solar carve out (as opposed to the entire RPS) and also have aggressive solar targets.

3. SREC Market Trends

3.1. Trading Volumes and Market Activity

Most SRECs Have Been Sourced from New Jersey; Other Markets Are Expanding

In recent years, the vast majority of SREC trading has occurred in New Jersey. Data on the source of SRECs traded or retired to meet state solar carve outs are available from the REC tracking systems used by the states to track RPS compliance.[6]

Trading and retirement data are generally similar for these markets, but SRECs could be traded multiple times before they are retired for compliance with a state solar carve out.

Figure 4 shows SRECs issued in 2010 in the PJM Interconnection balancing area, the most active region for SREC trading. PJM data provide a good indication of total SREC activity because most states with SREC policies are in or can sell into the PJM balancing authority, with notable exceptions being Massachusetts, New Hampshire, and to some extent North Carolina.[7]

In 2010, 75% of the SRECs issued in PJM were located in New Jersey, followed by Pennsylvania (11%), Ohio (6%), and Maryland (4%). Other states include SRECs from projects located in Virginia, North Carolina, Washington, D.C., and West Virginia.

Data through October 2011 show similar general trends, with Pennsylvania seeing a higher percentage of issued SRECs (17%).

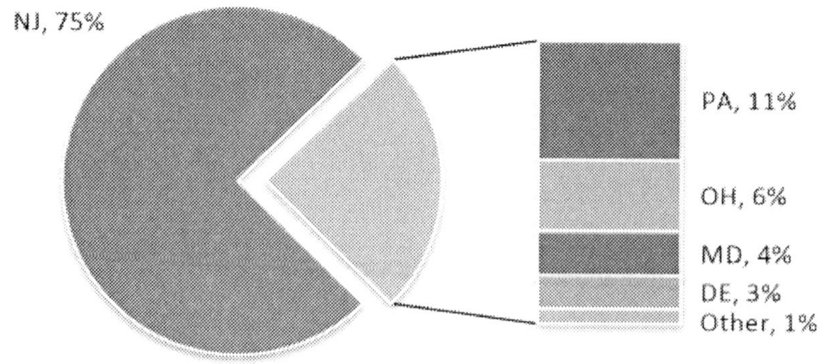

Source: PJM-GATS 2011a.

Figure 4. SRECs issued in PJM-GATS, 2010.

Solar capacity additions in recent years are consistent with the SREC trading data (see Table 8 and Figure 4). Of the states with SREC markets in 2010, New Jersey led installed PV by far, with 260 MW of cumulatively installed PV capacity, much of which came online in 2010, followed by Pennsylvania, North Carolina, and Massachusetts (Sherwood 2011). Note that North Carolina and Massachusetts are not included in the PJM-GATS SREC trading volume data. Missouri installations totaled less than 1 MW, but the Missouri solar requirement does not take effect until 2011.

Table 8. PV Capacity Additions (Annual and Cumulative)

State	2009 PV Additions MW_{DC}	2010 PV Additions MW_{DC}	Cumulative Installed Capacity MW_{DC}
DC	0.3	3.5	4.5
DE	1.4	2.4	5.6
MA	9.6	20.4	38.2
MD	4.7	3.4	10.9
MO	0.1	0.5	0.7
NC	6.6	28.7	40.0
NH	0.5	1.3	2.0
NJ	57.3	132.4	259.9
OH	0.6	18.7	20.7
PA	4.4	46.5	54.8

Source: Sherwood 2011

Notes: Figures here are calendar year additions. Capacity for each state represents in-state generation, although some states allow out-of-state capacity to contribute.

The dominance of SREC trades from New Jersey systems is not surprising because the New Jersey solar requirement was the first to take effect, its solar carve out is currently more than four times larger than the next largest state's requirement, and its interconnection requirements are such that only in-state resources can meet its standard. Further, New Jersey has had the highest SREC prices in the region. However, there has been a move toward greater geographic diversity of SREC markets. Monthly SREC trading data in PJM show an uptick in volumes traded in markets other than New Jersey in 2010 and 2011 (see Figure 5), particularly from systems in Maryland, Pennsylvania, and Ohio.

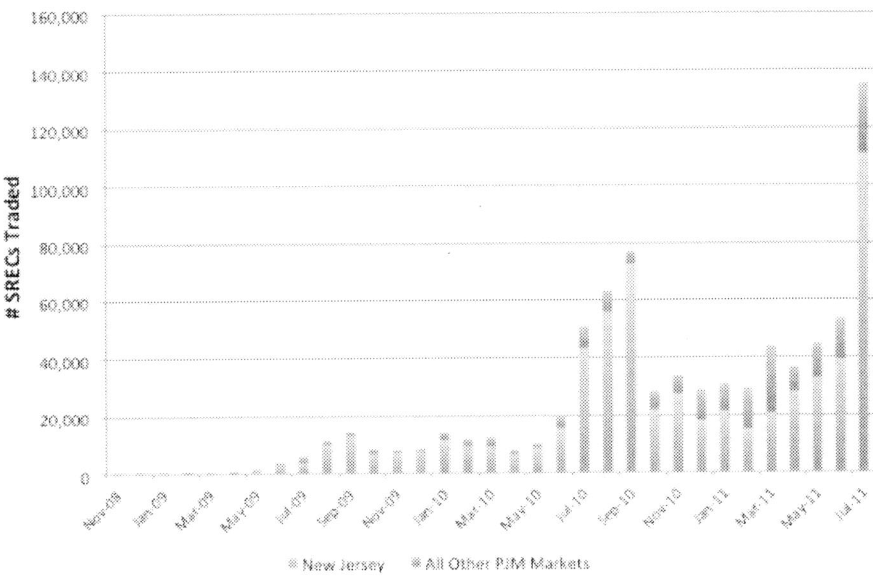

Figure 5. Monthly SREC trading volumes in PJM-GATS, 2009–2011.

SRECs can also be sold into voluntary markets in which individuals, businesses, and institutions make voluntary purchases of RECs to match their own electricity consumption. However, voluntary SREC transactions have been limited to date. In 2010, SRECs represented 0.2% of U.S. voluntary market sales, which includes sales by utility green pricing programs, competitive green power programs, and unbundled REC purchases (Heeter and Bird 2011). In some instances utilities use SRECs to include solar in the resource mix of utility green power programs offered to residential and small commercial customers, but they are a relatively small fraction of total program supplies. In 2010, utility programs procured about 39,000 MWh of solar energy to supply 0.8% of the green energy sold through their green pricing programs (Heeter and Bird 2011). In some instances, utilities have offered fixed-price, long-term contracts for the solar energy output and used the SRECs to supply green pricing programs. For example, the Tennessee Valley Authority (TVA) and North Carolina Green Power purchase the SRECs from PV system owners for a fixed fee of $0.10– $0.12/kWh for an established period such as 10 years and supply their green power programs with the SRECs (TVA 2011; NCGP 2011).

3.2. In-State Versus Out-of-State Sourcing

Several States are Sourcing SRECs Primarily from In-State Systems, While Others are Sourcing More Broadly

The location of solar projects providing SRECs is also highly dependent on rules regarding whether states can source SRECs from in-state or out-of-state facilities, as mentioned earlier. As a result of policies that allow regional sourcing, some developers are registering their projects in multiple states, particularly within the PJM balancing authority, to increase their opportunities to sell SRECs. However, the majority of systems are being used for compliance in state. Of the roughly 17,000 PV facilities registered in PJM-GATS,[8] 83% are registered for use in the SREC market of just one state, 12% are registered in two states, 5% are registered in three states, and 0.9% are registered in four states (PJM-GATS 2011b). Table 9 shows the number and aggregate capacity of PV systems registered for use in PJM-GATS states, as well as Massachusetts and North Carolina.[9]

Two states—New Jersey and Massachusetts—source all of their SRECs from in-state resources as required by state law. In another approach, Maryland regulations required electricity suppliers to seek SRECs first from generators on the Maryland distribution grid, meaning that suppliers could only procure SRECs from out of state if there were no available in-state resources. In 2010, Maryland electricity suppliers sourced 63% of the total SRECs retired from in-state generators, with about another quarter from Illinois, a state without a SREC market or solar carve out (see Figure 6a). In 2012, a binding in-state requirement goes into effect in Maryland (DSIRE 2011b).

Table 9. In-State Versus Out-of-State Project Registration as of October 2011

State	Registration	Number of Projects	Capacity (MWDC)
DC	In State	371	2.2
	Potential Imports	1,961	18.1
	Potential Exports	137	0.8
DE	In State	889	22.7
	Potential Imports	0	0.0
	Potential Exports	655	7.4
MA	In State	1,015	41.1
	Potential Imports	N/A	N/A
	Potential Exports	N/A	N/A
MD	In State	1,748	30.9
	Potential Imports	778	21.2
	Potential Exports	251	11.8
NJ	In State	10,170	417.0
	Potential Imports	N/A	N/A
	Potential Exports	35	5.7
NC	In State	125	63.3
	Potential Imports	80	41.2
	Potential Exports	75	1.7
OH	In State	417	27.5
	Potential Imports	2,752	57.9
	Potential Exports	144	1.3
PA	In State	4,873	110.7
	Potential Imports	1,503	37.7
	Potential Exports	2,683	58.6

Sources: PJM-GATS 2011b; Massachusetts DOER 2011b.

Note: "In state" refers to projects located within a particular state and registered only in that state. "Potential imports" refers to projects located in other states that are registered for potential use in the specified state. "Potential exports" refers to projects in the specified state that are registered for potential use to meet compliance obligations in other states. The majority of these potential export facilities are also registered in the state where they are located. As of August 1, 2011, the potential imports to Washington, D.C., decreased as a result of the District's decision to limit out-of-state generation.

In other markets, the source of SRECs retired for compliance requirements is more diverse. Washington, D.C.-based solar generators only provided 13% of SRECs retired to meet the D.C. standard in 2010. SRECs used for compliance in Washington, D.C., have come from 14 other states, including a number of states without SREC markets or solar carve outs, such as Virginia (see Figure 6b). In Washington, D.C., external facilities are five times more numerous than facilities located within the District (see Table 9). Although Washington, D.C.'s previous law stated a preference for in-district resources, its newly adopted law (August 2011) has a more stringent local requirement, starting in 2011. Out-of-state projects certified for the RPS program by January 31, 2011, will be eligible to continue to sell into the Washington, D.C., market.

While Pennsylvania has a substantial amount of registered in-state facilities, only about one-third of SRECs supplying Pennsylvania's solar requirement were sourced from in-state facilities in 2010. The remainder was sourced from six other states, with the largest contributors being Delaware (41%) and Illinois (25%) (see Figure 7a). A small fraction of SRECs have been retired from systems located in a variety of states within PJM to meet the Pennsylvania solar standard (see Figure 7a). Pennsylvania systems are also being retired for compliance with requirements in Washington, D.C., and Ohio.

Despite the 50% in-state requirement in Ohio, only 35% of SRECs retired in 2010 were sourced from Ohio (see Figure 7b). This may be the case because Ohio utilities and competitive providers did not achieve full compliance with the unadjusted solar carve outs in 2010; several entities successfully petitioned to have their 2010 targets adjusted downward, with any shortfall to be made up in 2011 (Siegfried 2011). In Ohio, there are about six times as many out-of-state facilities registered (see Table 9). Pennsylvania contributed 63% of the total SRECs retired to date for compliance with the Ohio solar target.

In North Carolina, the utilities are sourcing SRECs from both out-of-state projects, which are limited to 25% of the standard, as well as from in-state sources. Duke Energy Carolinas maintains a diverse solar portfolio, including: Duke-owned PV installations (at its central sites and at distributed sites[10]), credits purchased from out-of-state suppliers (which can be used for up to 25% of compliance requirements), credits purchased from in-state PV and solar thermal projects, and credits purchased from a standard offer contract program available to the residential and commercial sectors (Duke Energy 2011b). Similarly, Progress Energy Carolinas purchases SRECs from in-state and out-of-state generators, as well as through the utility's standard offer contract program (Progress Energy Carolinas 2011a).

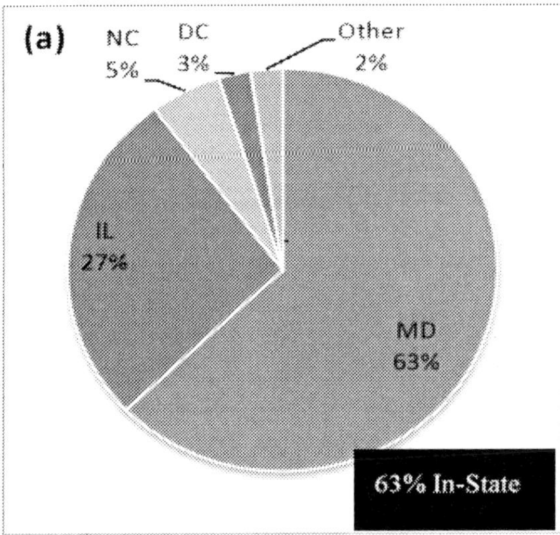

(a) Other includes: DE, IN, NY, OH, PA, VA, WV.

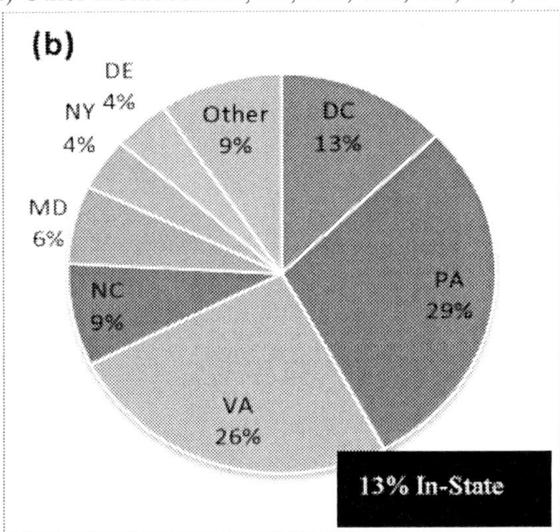

(b) Other includes: IL, IN, KY, MI, NJ, OH, WI, WV.

Source: PJM-GATS 2011a.

Figure 6. (a) Source of SRECs retired for 2010 compliance in Maryland (b) Source of SRECs retired for 2010 compliance in Washington, D.C.

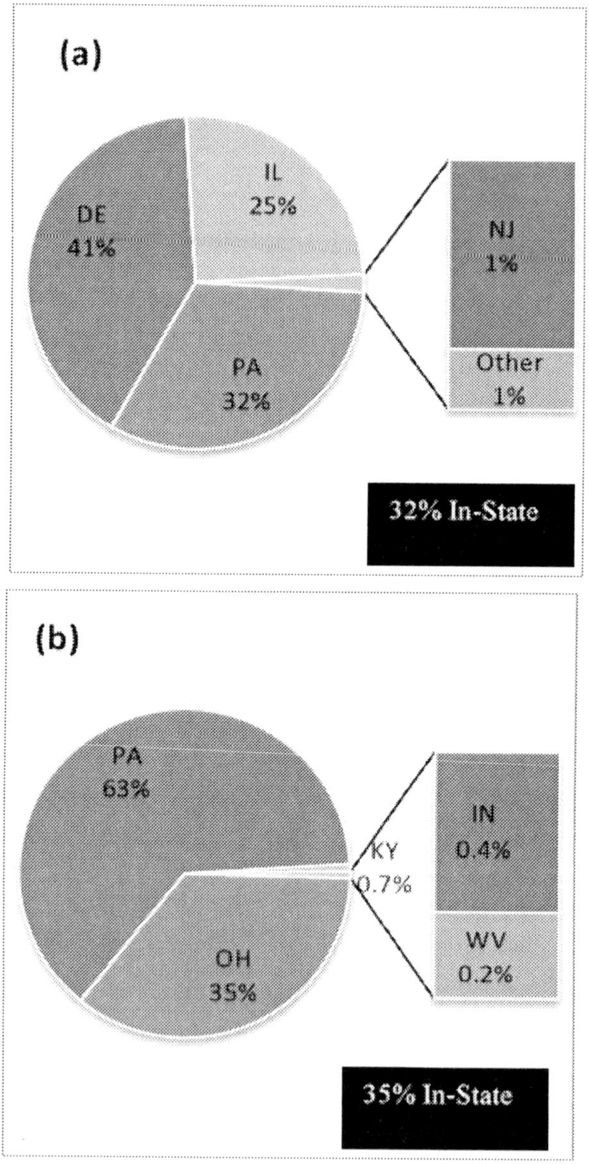

Source: PJM-GATS 2011a.
(a) Other includes: MD, VA, WV.

Figure 7. (a) Source of SRECs retired for 2010 compliance in Pennsylvania (b) Source of SRECs retired for 2010 compliance in Ohio.

In Missouri, the three investor-owned utilities (IOUs) plan to procure most of their SRECs from solar projects in the West that were registered in WREGIS, according to plans filed in 2011 (Ameren 2011b; KCP&L 2011; KCP&L-GMO 2011). Additionally, Ameren plans to source some SRECs both from a 100 kW installation on their company headquarters and some through entering into standard offer contracts with residential and commercial customers (Ameren 2011b).

For the 2011–2013 planning period, Ameren Missouri plans to procure SRECs from out-of-state sources, through a continuation of their standard offer program, potentially through utility-scale installations at several of its existing generating facilities and through unsolicited proposals from solar developers (Ameren 2011b). The other two utilities plan to continue to meet the solar requirement by purchasing SRECs from out-of-state projects through brokers due to price considerations and uncertainty regarding the solar carve out and RPS (KCP&L 2011; KCP&L-GMO 2011).

3.3. Size of Systems Supplying SRECs

SREC Markets are Supporting a Mix of PV System Sizes; Large Systems are Increasingly Common

Policymakers may be interested in ensuring that SREC markets support a mix of system sizes in order to: (1) increase the supply of decentralized, local power generation at customer sites; (2) establish support for a variety of contractors, installers, and developers; and/or (3) ensure that a variety of customer types benefit from ratepayer-funded programs.

These goals may also be balanced with an interest in achieving solar targets at least cost.

To date, most markets are supporting a mix of residential and mid- to large-sized commercial PV systems, but large systems are increasingly dominating capacity in some markets (see Figure 8).

In all markets except North Carolina, approximately two-thirds or more of the total number of registered projects are smaller than 10 kW (see Figure 8), with mid-sized (10 kW to 250 kW) projects generally contributing substantially to the total project number.

In North Carolina, about 15% of registered projects are smaller than 10 kW, nearly 50% are midsized, and the remainder is large projects that are greater than 250 kW in size.

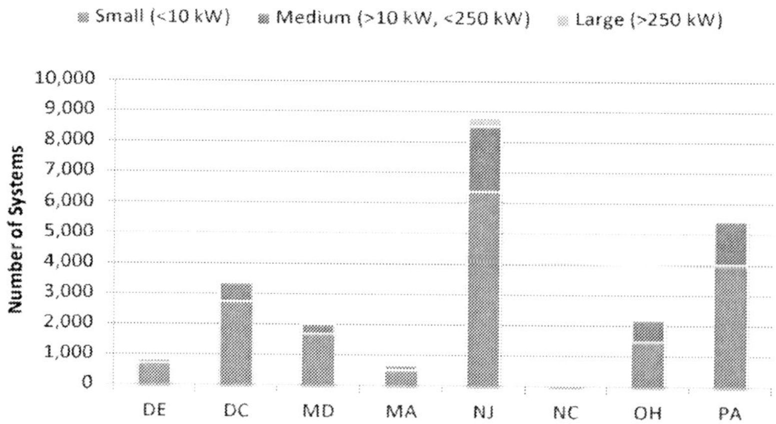

Sources: Project registration data. PJM-GATS 2011b; Massachusetts DOER 2011b; N.C. Utility Commission 2011a.

Figure 8. Number of PV projects by size and jurisdiction.

Figure 9 shows the relative contribution of small (<10 kW), mid-sized (10 kW to <250 kW), and large systems (>250 kW) to total registered solar capacity. Large projects currently dominate registered capacity in a few markets—most notably, North Carolina, Massachusetts, and New Jersey. For example, although nearly 75% of the 8,800 projects in New Jersey are smaller than 10 kW, they represent 12% of the aggregate capacity. In North Carolina, projects greater than 250 kW represent more than 90% of the aggregate capacity, while small projects account for less than 1%.

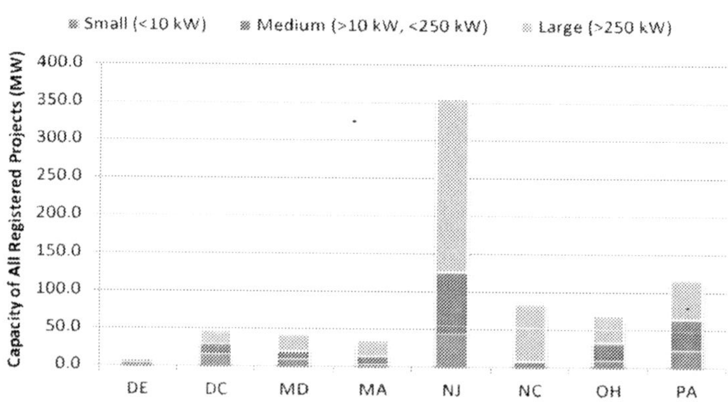

Sources: PJM-GATS 2011b; Massachusetts DOER 2011b; NCUC 2011a.

Figure 9. Aggregate capacity (in MW) of registered PV projects by size and jurisdiction.

Small residential and commercial systems may inherently face additional barriers in SREC markets. Small systems may not be as economic as larger systems due to the lack of efficiencies of scale, although this may be offset by the fact that residential systems displace higher retail electricity rates. Consumers may have a difficult time finding buyers for their SRECs; buyers are often interested in purchasing large quantities at once to minimize transaction costs. In addition, small consumers may not be able to access complex incentive programs, where they are required to bid into a competitive process. Lastly, small system owners may not want to take on the risk of uncertain future SREC revenues.

Available data suggest that these challenges have not prohibited the development of small systems in most states to date, however. In most markets, SREC aggregators are helping small residential and small commercial projects sell their SRECs. For example, in Massachusetts, 99.8% of small-sized projects and 91% of mid-sized projects were part of an aggregation, as of July 2011 (Massachusetts DOER 2011b). In the rest of PJM, 55% of small systems and 40% of mid-sized projects participate in aggregations (Bowery 2011). Often utilities have used standard offers, auctions, and solicitations to help encourage development of projects in various size categories. Under these programs, utilities purchase the SRECs under long-term contracts, which provide a stable incentive for system owners. In Missouri, North Carolina, and Ohio, regulated utilities have created standard offer programs for their retail customers who are interested in installing behind-the-meter systems on their homes or businesses (see text box in Section 2.6).

Additionally, separate from standard offer programs, other LSEs have solicited long-term SREC contracts from developers of both small and large projects through competitive solicitations and auctions. Since most LSEs in an unregulated market do not want to sign a contract beyond the scope of their demand requirement, competitive solicitation and auction programs have been an important way for independent power producers to obtain long-term contracts. The New Jersey SREC-Based Financing Program (see Appendix) solicits SRECs from projects up to 2 $_{\text{MWDC}}$ in size. The program has a small project size segment (<50 kW) and a large project size segment (50–2,000 kW). Also, Pennsylvania utility PPL Electric Utilities has chosen to procure SRECs from projects <10 kW through a "Small Scale Solar Set-Aside" RFP targeted at SREC aggregators (Appendix). These programs have been important for encouraging a mix of system sizes in these states.

System size has been trending larger in recent years. There has been a steady increase in the number of PV projects larger than 250 kW. The number of systems greater than 250 kW has grown cumulatively from just two in 2003 to nearly 400 through the first half of 2011. Since 2008, the average size of projects in this class has been greater than 600 kW. As expected, the average has been skewed by the installation of some large utility-scale projects in recent years. For example, as shown in Figure 10, the average size of this class of projects was 1.1 MW in 2009, which is much higher than previous years due to the completion of several utility-scale projects.

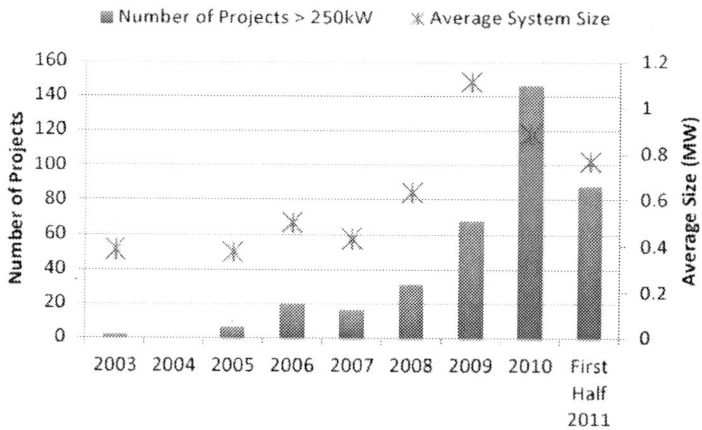

Source: PJM-GATS 2011b.

Figure 10. Number and average size of facilities >250 kW installed annually in PJM-GATS.

Utility-scale PV projects >1 MW have played an increasingly larger role in the markets in recent years. New Jersey leads in utility-scale solar deployment with 51 solar projects >1 MW operational as of August 2011 (PJM-GATS 2011b). North Carolina currently has 22 in-state generators >1 MW completed (NCUC 2011a). Massachusetts has three operational PV generators >1 MW (Massachusetts DOER 2011b), and Delaware is home to a 10 MW solar park commissioned this year (SEIA 2011).

A number of large projects are planned over the next few years. In aggregate, these planned projects could result an estimated 400 MW of new PV deployment. There is variation in estimates of planned capacity and some projects may not come to fruition, but these data are indicative. Figure 11 shows the aggregate capacity of PV projects completed versus planned as of September 2011.

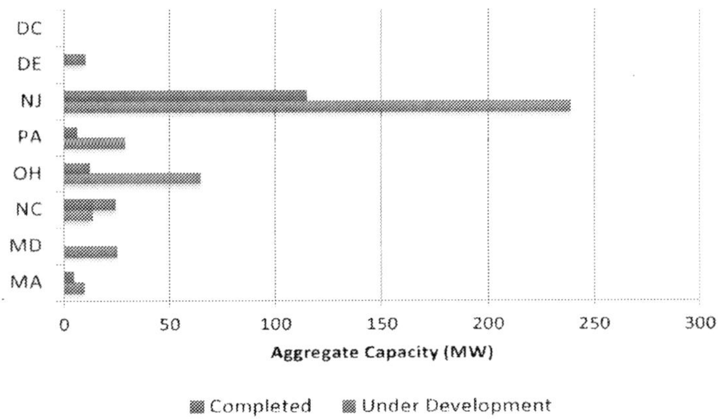

Sources: PJM-GATS 2011a; Massachusetts DOER 2011b; NCUC 2011a; SEIA 2011; SEPA 2011; NJ BPU 2011c.

Note: The New Jersey estimate for projects under development is derived from the list of projects that have registered with the NJ BPU under the SRP program prior to construction but have not yet been completed.

Figure 11. Aggregate capacity of utility scale (>1 MW) PV projects (completed and under development).

Large projects can help states achieve compliance more quickly. However, they may flood the market with excess credits, which could reduce SREC prices, potentially to levels that could no longer provide adequate incentives for small PV systems. In Ohio, two utility-scale PV facilities are planned for the near term. Together, if completed, the 12-MW Wyandot Solar Facility and the 50-MW Turning Point facility would produce an estimated 24% of the total Ohio SREC requirement in 2013, according to an estimate by SRECTrade (SRECTrade 2011b). Two main mechanisms are used by states to limit the impact of large systems on the market. The first mechanism is to place a cap on the size of systems eligible to sell SRECs—2 of the 10 jurisdictions currently have such a cap. In Massachusetts, solar generators must be smaller than 6 $_{\text{MWDC}}$ in order to generate SRECs. Washington, D.C., adopted legislation effective August 1, 2011, which caps systems eligible to generate SRECs at 5 MW. Pennsylvania has considered capping system size. Before 2010, New Jersey had a 2-MW DC cap, which was lifted to help keep pace with aggressive solar growth targets.[11] The other mechanism is to establish separate standards for customer-sited and large-scale systems. For example, Colorado (which does not have a tradable SREC market) requires half of the solar carve out to come from customer-sited systems (DSIRE 2011a).

3.4. SREC Prices

New Jersey has Historically Experienced the Highest Prices; Recent or Forward Prices Have Dropped Significantly in Most Markets

SREC prices tend to track the supply and demand in the state or regional market. Spot pricing for SRECs is publically available from platforms like SRECTrade and FlettExchange.[12] SRECTrade hosts a monthly auction, while Flett Exchange is an online exchange. Both platforms cover markets in PJM states, Massachusetts, and Ohio, and similar price trends can be seen in reported data from both companies. Figure 12 shows SREC prices for the current or nearest compliance year. Historically, SRECs from New Jersey have seen the highest prices, ranging from \$400/MWh to approximately \$650/MWh in recent years, while SRECs from most other states have ranged from \$200/MWh to \$500/MWh (Figure 12). However, in 2011, SREC prices fell, sometimes dramatically, in all markets except for in-state Massachusetts and Ohio. In New Jersey, spot prices fell from between \$600/MWh and \$700/MWh in 2009 and 2010 to less than \$200/MWh in September 2011 (Figure 12).

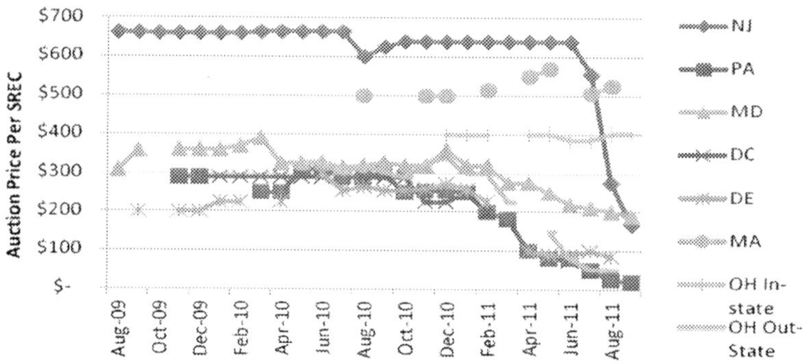

Source: SRECTrade 2011c.

Figure 12. SREC spot prices (per MWh), August 2009 to September 2011.

Long-term contract prices can differ substantially from spot prices. Solicitations held by EDCs in New Jersey for procuring SRECs in 10- or 15-year contracts have resulted in SREC prices lower than spot prices in 2009 and 2010. In February 2011 and June 2011, average 10-year SREC contract prices ranged from $280/MWh to $448/MWh, depending on the project size and timing of the solicitation (Table 10).

Table 10. New Jersey EDC Solicitation 10-Year Contract Pricing, 2011

RPS Reporting Year 2011	Solicitation 6			Solicitation 7		
	Small	Mid-Sized	Large	Small	Mid-Sized	Large
Date bids submitted	2/17/11	2/17/11	2/17/11	6/10/11	6/10/11	6/10/11
Number of bids submit (participation)	86	78	27	83	115	29
Contracts awarded	59	35	12	35	19	9
Total capacity awarded (kW)	778	7,557	8,231	693	9,170	9,512
Average size (kW)	13	216	686	20	483	1,057
Average Net Present Value for contracts ($/kW)	$3,166*		$2,926	$2,683	$2,145	$1,983
Corresponding 10-yr renewable energy credit contract price ($/MWh)	$448*		$414	$379	$303	$280
Lowest Net Present Value ($/kW)	$2,819*		$2,424	$2,248	$1,759	$1,774
Corresponding 10-yr renewable energy credit contract price ($/MWh)	$399*		$343	$318	$249	$251

*In earlier solicitations, small and mid-sized projects were not solicited separately. The BPU's Solicitation 6 report does not distinguish between average SREC price for small and mid-sized projects. Source: NJ BPU 2011a; NJ BPU 2011b.

In Pennsylvania, SREC spot prices dropped to less than $50/MWh in September 2011 from approximately $300/MWh in mid-2010 (FlettExchange 2011), presumably due to oversupply in the market. Long-term (8.5 to 10.0 years) SREC contracts held by EDCs in Pennsylvania have seen prices ranging from $149/MWh to $286/MWh, as evidenced by prices from contracts solicited in February, May, and June 2011 (Table 11).

Table 11. Pennsylvania EDC Solicitation Contract Pricing, 2010–2011

EDC	Date of Solicitation(s)	Delivery Period	Total SREC Volume	Average Price/ SREC
Met-Ed/Penelec	Feb & June 2010	10 years	Feb: 10,000, June: 15,000	$286/MWh
Penn Power	Feb 2011	9 years	19,800	$199/MWh
PPL Electric	May 2011	8.5 years	25,500	$149/MWh
PPL Electric	June 2011	9 years	1,000	$149/MWh*

Sources: First Energy Corporation 2011a; First Energy Corporation 2011b; PPL Electric 2011a; PPL Electric 2011b.
*Pricing was determined by the results of the May 2011 solicitation.

In Washington, D.C., prices have declined as SRECs from other states have saturated the D.C. market. In 2010, SREC spot prices were closing between $200/MWh and $300/MWh, whereas in mid-2011, SREC spot prices dropped to between $50/MWh and $80/MWh. The Council of the District of Columbia addressed these issues by closing the door to most new out-of-district resources (current out-of-district resources will be grandfathered in) and increasing the ultimate solar requirement from 0.4% to 2.5% by 2023.

3.5. Compliance in SREC Markets

Compliance has Proved Challenging in Some SREC Markets, but Improved Compliance is Expected in Some Markets as Indicated by Price Drops and Installation Trends

The relatively high SREC prices in recent years, in many cases near the levels of solar ACPs, reflect the fact that some states have struggled with solar carve out compliance. However, the substantial drop in forward prices of SRECs in most states indicates that compliance challenges are dissipating and that markets may be oversupplied in some instances.

The largest shortfalls historically have been seen in New Jersey, which has by far the largest requirement. New Jersey experienced a large shortfall in 2009, with utilities paying about $39 million in solar ACPs (Figure 13). In 2010, the shortfall was slightly less, though still substantial, at $33 million. Recently, solar installations in New Jersey have surged, indicating that New Jersey will see fewer shortages in upcoming compliance periods.

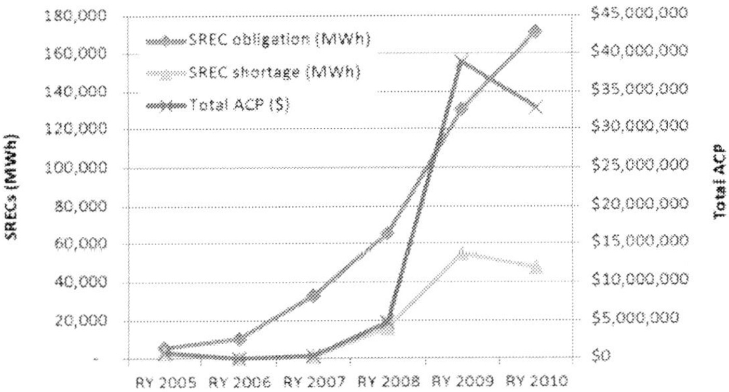

Note: RY stands for reporting year.

Figure 13. New Jersey solar carve out compliance.

Maryland, Ohio, and Washington, D.C., have experienced shortages in SREC compliance in recent years, though compliance is looking more likely for 2010 and 2011 based on price trends in these markets. In Washington, D.C., future compliance is more difficult to predict, given the recent change in the compliance requirement. Table 12 shows compliance shortages and solar ACP payments in states with SREC targets in 2009 or earlier. In Pennsylvania, compliance has been nearly 100%, with one supplier choosing to pay the ACP for two SRECs in 2008. Shortages are unlikely in the future due to the pace of installations in the state.

Maryland experienced shortfalls in 2008 and 2009, which was largely due to a shortage in supply. The 2008 Maryland SREC requirement was approximately 3,000, but only 276 SRECs were generated in 2008. Similarly, approximately 6,000 SRECs were needed to meet Maryland's 2009 requirement, but only 3,495 were generated. In 2008, Maryland utilities were short by 2,707 MWh and paid approximately $1,218,000 in solar ACPs. The shortage continued in 2009 when Maryland utilities were short by 2,865 MWh and paid $1,147,000 in solar ACPs.

In Washington, D.C., utilities did not submit any SRECs for compliance in 2007 or 2008. This resulted in shortages of 592 MWh in 2007 and 1,255 MWh in 2008, requiring solar ACPs totaling $178,000 and $377,000, respectively. In 2009, utilities continued to be short, this time by 851 MWh, resulting in $426,000 in solar ACPs.

Table 12. State Solar Carve Out Compliance, 2007–2009

	2007			2008			2009		
	SRECs required (MWh)	SREC shortage (MWh)	Total solar ACP ($1,000)	SRECs required (MWh)	SREC shortage (MWh)	Total solar ACP ($1,000)	SRECs required (MWh)	SREC shortage (MWh)	Total solar ACP ($1,000)
DC	N/A	N/A	N/A	900	146	$36	1,091	65	$18
DE	592	592	$178	1,225	1,225	$377	2,235	851	$426
MD	N/A	N/A	N/A	2,934	2,707	$1,218	6,125	2,865	$1,147
OH	N/A	N/A	N/A	N/A	N/A	N/A	3,955	2,858*	None**
PA	26	0	0	348	2	Not reported	1,221	-	None

*The total number of SRECs retired in Ohio is unknown because compliance reports by Competitive Retail Electric Service are deemed confidential. The figures here represent only the publicly available information provided by Ohio's Electric Distribution Utilities. The Ohio PUC estimated that 5,598 SRECs would be required in 2009 for the entire market.

**The Ohio PUC approved force majeure petitions by utilities, did not assess financial penalties, and added 2009 shortages to the 2010 obligation.

Note: Massachusetts and North Carolina began compliance in 2010. Missouri does not have a compliance obligation until 2011.

In Ohio, utilities and competitive suppliers were unable to meet the in-state solar carve out requirement in 2009 but successfully petitioned to have the in-state solar requirement reduced for that year due to *force majeure*. In Ohio, the RPS includes a *force majeure* provision that enables obligated entities to request commission action in the event of an unforeseen circumstance outside of their control.

In Ohio, utilities and competitive providers have stated that they made efforts to secure in-state solar SRECs but were not able to obtain enough SRECs through that process. FirstEnergy, the parent company of three Ohio utilities, also offered a residential SREC purchase program but was only able to procure 13 SRECs through the program in 2009 (Ohio Edison Company et al. 2010). Preliminary compliance reports have been issued in Ohio, while some utilities and competitive providers filed for *force majeure* in 2009. In 2010, most of the filers for *force majeure* were competitive providers and represented smaller volumes than 2009 (Siegfried 2011).

New Hampshire's first compliance year for its solar carve out was 2010. In 2010, entities paid $58,884 in ACPs (NH PUC 2011). At a solar ACP level of $160/MWh for 2010, the payment represents a shortage of 368 SRECs, or 8% of the total obligation of 4,255 SRECs (Epsen 2011).

Data from Massachusetts, Missouri, and North Carolina are not yet available because North Carolina and Massachusetts began compliance in 2010, and Missouri's first compliance year is 2011. Preliminary data indicate that a large shortage exists in Massachusetts (which had its first compliance year in 2010).

Data from NEPOOL GIS indicate that only 2,738 MWh of qualifying Massachusetts solar were issued in 2010, compared to a requirement of approximately 34,000 MWh (Massachusetts DOER 2011c). The gap between installations and the requirement helps explain why Massachusetts SREC prices have not fallen as they have in other state markets.

Missouri's first compliance year is 2011 but is not expected to see shortages due to the ability of utilities to obtain SRECs from the national market. Similarly, in North Carolina, which had its first compliance year in 2010, utilities are expected to meet their solar requirements. SRECs can be banked for an unlimited amount of time in North Carolina and 25% of the SRECs can be procured nationally.

One large IOU, Duke Energy, projects that it has contracts in place for a sufficient number of SRECs to meet its solar requirements through 2018 (NCUC 2011b).

4. Future Outlook

Looking forward, SREC markets are expected to continue to expand as state solar requirements ramp up over time from more than 520 MWAC in 2011 to approximately 2,000 MWAC by 2015 and 4,300 MWAC by 2020.

While several states struggled to achieve compliance recently, some markets show signs of improvement. A sharp drop in SREC prices in several states indicates that the markets may be oversupplied and on track to achieve near-term compliance targets. For example, energy year 2012 SREC spot pricing in New Jersey is at about $150/MWh compared to energy year 2011 spot market prices of more than $600/MWh. SREC prices have also eased in Pennsylvania. Other markets, such as Massachusetts, may continue to face shortages in the near term as its standard ramps up.

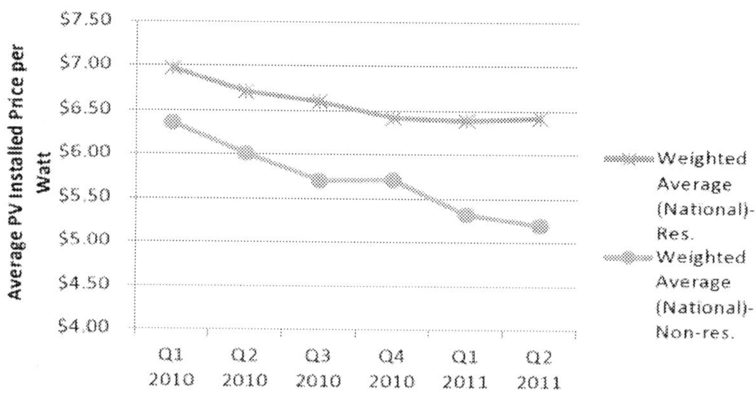

Sources: SEIA GTM 2011a; SEIA GTM 2011b.

Figure 14. Average PV installed price Q1 2010 through Q2 2011.

Substantial increases in installed capacity have been fueled by a 30% drop in module prices in the last year as well as slower growth in international markets that have placed greater emphasis on U.S. markets. While module prices declined from $2.21/W in the first quarter (Q1) of 2010 to $1.56/W, in Q2 2011, installed costs have not fully reflected these cost reductions yet, particularly in the residential market segment (SEIA GTM 2011a; SEIA GTM 2011b). Solar Energy Industries Association and Greentech Media report an approximate $0.50/W drop in residential installed PV costs nationally from the beginning of 2010 through Q2 2011 and more than $1.15/W drop in non-residential installed costs nationally over the same period (see Figure 14).

Going forward, module cost reductions are likely to contribute to additional drops in installed costs. Given the drop in SREC prices in a number of markets, there will be increased pressure to drop installation prices to offset reduced revenue streams (SEIA GTM 2011b).

The current pace of project development is also influenced by the rapidly approaching expiration date of the Federal Treasury 1603 cash grant program, currently set to expire at the end of 2011. For projects to be eligible for the cash grants, they must meet safe harbor rules for commencing construction before the end of 2011, although completion is not required until 2016. Once the cash grant program expires, large projects in particular may require tax equity investors in order for the project to monetize the investment tax credit. The rate of development going forward will be influenced by the availability of tax equity and whether it will keep up with project demand. Also, the elimination of the cash grants may mean that developers place even greater emphasis on long-term contracts for SRECs to obtain debt financing for projects; thus, the continued availability of long-term contracting provisions or price floors will continue to be important.

Going forward, the increase in utility-scale PV projects may impact the supply and demand balance in state SREC markets, particularly those with relatively small carve outs. Large projects have the potential to achieve significant portions of the solar targets, which could result in rapid reductions in SREC prices. These effects may be particularly prevalent in some markets where large-scale projects have been planned, such as New Jersey, Ohio, and Maryland.

In New Jersey, the substantial uptick in installations in 2010 and the first half of 2011 are expected to keep the market in oversupply in the near term. Figure 15 shows the monthly trends in PV capacity installations in New Jersey. SRECTrade estimates that the average capacity required to meet the 2012 target is 370 MW, meaning that over the 2012 energy year, an average of 370 MW would need to be operating (SRECTrade 2011d). As of August 31, 2011, New Jersey had 430 MW installed (NJ OCE 2011). In addition to the current capacity expansion, the number of projects registered in New Jersey increased from just over 900 approved registrations at the end of August 2010 to over 3,700 approved projects by the end of July 2011 (representing over 430 MW) (NJ BPU 2011c).[13] While not all of these projects will ultimately be built, the backlog in registrations indicates that there is a potential for oversupply going forward in the New Jersey SREC market. As a result of recent installations and the existing project queue, industry analysts expect that the New Jersey market will be oversupplied through 2012 but may pick up again in 2013 (SEIA GTM 2011b).

Maryland had 42 MW installed at the end of Q2 2011 and 10.8 MW installed in the first half of 2011 alone, which industry analysts attribute to the newly available residential solar lease (SEIA GTM 2011b). Residential solar rebates are also available from the state. At least two large projects (>15 MW) are expected to begin commercial operation in Maryland in 2012, which would also play a substantial role in helping the state meet its compliance targets. However, it is unclear how the Maryland market will react when it is closed to out-of-state SRECs in 2012, but this will provide a boost for in-state projects.

In Pennsylvania, installations remained strong in the first half of 2011 due to significant project backlogs, despite an earlier drop in SREC prices. As of the first half of 2011, 105 MW had been installed, which is well ahead of current targets (SEIA GTM 2011b).

Declining PV installation costs may have played a role in the uptick in installations; from Q2 2010 to Q2 2011, residential installation prices in Pennsylvania have declined from \$7.13/W to \$5.94/W, and non-residential installation prices have decreased over the same period from \$6.23/W to \$5.17/W (SEIA GTM 2011b). The market is expected to slow in the near term but could recover in 2013, perhaps in part to serve markets such as Illinois (SEIA GTM 2011b).

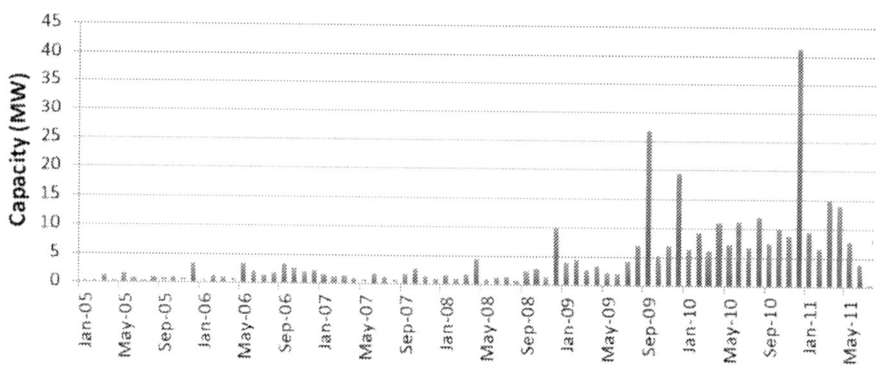

Source: PJM-GATS 2011b.

Figure 15. Capacity installed by month in New Jersey.

Recent legislative changes will have an impact on markets going forward. While most recent changes have resulted in stronger SREC targets, the uncertainty created by evolving policies can be a barrier to project development. Policymakers in Washington, D.C., and Connecticut have strengthened solar polices, while Massachusetts faced some criticism by solar developers for decreasing the solar ACP. In August 2011, Washington, D.C., policymakers expanded the solar requirement from 0.4% by 2020 to 2.5% by 2023 and specified that it be met by systems no larger than 5 MW and generally sited within D.C. In Connecticut, legislation (Bill 1243) was recently adopted requiring distribution companies to procure RECs from facilities <1 MW, spending $8 million in the first year and increasing by that same amount annually to $32 million in the fourth year. The companies must solicit 15-year contracts from small, mid-sized, and large facilities up to 1 MW. Finally, Massachusetts recently lowered its solar ACP from $600/MWh to $550/MWh and has proposed a long-term schedule reducing solar ACPs by 5% annually from 2013 to 2021, which helps provide certainty to the market for forward pricing.

Other state legislatures have recently considered bills to expand state solar carve outs, although none of these provisions have passed to date. In New Jersey, the Senate passed S2371, which would increase the SREC requirement in 2013 from 596 GWh to 772 GWh (the current 2014 requirement), with each subsequent annual requirement moving forward one year. In Pennsylvania, amendments (HB 1580) to the original RPS Act were introduced that would restrict solar eligibility to in-state resources and would increase the requirements for 2013–2015 compliance years from approximately 70 MW in 2013 to 207 MW in 2015 (SRECTrade 2011e). The New York legislature considered a bill this year to establish a solar carve out with an initial compliance requirement of 0.15% in 2012.

CONCLUSION

SREC markets have been implemented by a number of states to help encourage the development of solar energy systems and to facilitate compliance with state RPS solar carve outs. These markets are relatively new with some markets just beginning to operate while others have had several years of implementation experience. This report documents early experience with the design and implementation of SRECs markets. The key findings of this report are the following:

- ***SREC markets are expected to grow rapidly in coming years as state solar requirements ramp up.*** Of the 10 jurisdictions that allow and anticipate the use of SREC trading, the solar carve outs are scheduled to grow from more than 520 MWAC in 2011 to nearly 7,300 MWAC in 2025. Targets for solar generation vary from 0.2% to 3.5% of retail electric sales. SREC markets are dominated by PV, although solar thermal is eligible to meet solar requirements in a number of states.

- ***Several states limit eligibility to in-state development, while a few allow SRECs from a broader geographic region.*** Two jurisdictions—Washington, D.C., and Maryland—recently strengthened their in-state requirements, which will affect future development. A small national market for SRECs has resulted from solar requirements in North Carolina and Missouri that allow SRECs to be sourced from anywhere in the United States.
- ***Solar ACPs are scheduled to decline over time reflecting expectations of declining PV costs.*** Solar ACPs can be significant because they reflect the ceiling on SREC prices, and in shortage situations, SRECs typically trade near the solar ACP. New Jersey has had the highest solar ACP initially set at $700/MWh, but it is scheduled to be reduced to $600/MWh by 2015. Other state solar ACPs are scheduled to range from $300–$500/MWh by 2015.
- ***New Jersey has been the dominant SREC market to date, but SREC markets in other states are expanding.*** New Jersey has dominated because it has the earliest and largest solar requirement (more than four times larger than any other state). Additionally, the high New Jersey solar ACP has driven SREC prices to be the highest in the nation in recent years. However, SREC requirements in other states are growing, and projects are being implemented in a broader region. Voluntary markets for SRECs have been very limited given the substantial price differential between SRECs and RECs sourced from other technologies.
- ***SREC spot prices have eased recently in a number of markets as installations have surged and compliance challenges have eased.*** SREC prices have ranged from about $400–$650/MWh in New Jersey and about $200–$500/MWh in other states in recent years. In August 2011, spot prices for energy year 2012 SRECs in New Jersey dropped to $150/MWh compared to $500/MWh in May 2011. Pennsylvania energy year 2011 spot prices fell to about $50/MWh in August 2011 as a result of oversupply in the market. Similarly, in Washington, D.C., SREC spot prices dropped to $50–$80/MWh in summer 2011 from $200–$300/MWh in 2010 because out-of-district SRECs were

allowed. In response, Washington, D.C., adopted legislation generally restricting eligibility to in-district systems and expanded the solar targets substantially to reduce excess supply. A number of other states have introduced legislation that would expand the markets, but to date, no other legislation has passed.

- ***SREC markets are supporting a mix of PV system sizes, including residential and small to large commercial systems.*** In most markets, about two-thirds of registered projects are <10 kW. Projects >250 kW dominate capacity installed in a few markets, including North Carolina, Massachusetts, and New Jersey. There has been a trend toward large-sized projects recently, with the average size system in PJM exceeding 750 kW in the last few years.
- ***Installation of large systems could significantly impact compliance levels and prices.*** Large projects planned for New Jersey, Ohio, Maryland, and perhaps other states could impact prices and compliance levels. Only Massachusetts and Washington, D.C., currently cap the size of systems that can generate SRECs.
- ***Compliance has proved challenging in a number of SREC markets in recent years, but adequate supplies are expected going forward in many of these regions.*** A handful of states reported compliance shortfalls in 2009. The first compliance periods for a few markets occurred in 2010 or 2011. However, compliance challenges have eased in a number of markets recently, as evidenced by substantial declines in SREC prices. Surges in the rates of installations in New Jersey and Pennsylvania in particular have placed those states ahead of targets.
- ***Lack of long-term contracts has been a barrier to project developers obtaining financing in some markets.*** Several states have instituted long-term contracting requirements, price floors, or other provisions to try to overcome these challenges. The New Jersey program in particular has led to a greater number of long-term contracts, which has facilitated project development. However, the drop in SREC prices may mean that ratepayers must make up the difference between spot and contract prices. The price floor mechanism in Massachusetts has not been fully tested, but investors appear to be gaining comfort with this mechanism. The need for long-term contracting may continue to pose a problem, particularly in areas where rebates or other supplemental incentives expire and SRECs become the primary incentive stream for financing systems.

- ***Rate caps exist in some form in five SREC markets but have not been reached or are yet to be evaluated.*** The modest size of most solar carve outs suggests that it is not likely that rate caps will be reached in the near term.

References

AEP Ohio. (2011). Ohio Power Company Renewable Energy Technology Program Rider, PUCO #19. Issued June 22. https://www.aepohio.com/global/utilities/lib/docs/save/renewable/RETP_Rider.pdf.Accessed August 15, 2011.

Ameren Missouri. (2011a). "Standard Offer Contract." http://www.ameren.com/sites/aue/Environment/Renewables/Pages/StandardOfferContract.aspx. Accessed August 10, 2011.

Ameren Missouri. (2011b). "Renewable Energy Standard Compliance Plan 2011–2013." https://www.efis.psc.mo.gov/mpsc/. Accessed August 15, 2011.

Barbose, G. (October 31, 2011). Email. Lawrence Berkeley National Laboratory, Berkeley, CA.

Bernstein, H.; Judge, M.; Andrews, N. (14 July 2011). Personal communication. Massachusetts PUC and Massachusetts Department of Energy Resources, Boston, MA.

Bingaman, B. (2010). "Re: Policy Statement in Support of Pennsylvania Solar Projects Docket No. M-2009-2140263." www.puc.state.pa.us/pcdocs/1070428.pdf. Accessed August 3, 2011.

Bowery, B. (3 November 2011). Personal communication. SREC Trade, San Francisco, CA.

California Public Utilities Commission (CPUC). (7 October 2011). "Decision Implementing Portfolio Content Categories for the Renewables Portfolio Standard Program." Proposed decision 11-05-005. http://docs.cpuc.ca.gov/efile/PD/144972.pdf. Accessed November 3, 2011.

CPUC. (2011). "California Solar Initiative (CSI) Progress Report." Q2, Data Annex. http://www.cpuc.ca.gov/PUC/energy/Solar/legreports.htm. Accessed September 30, 2011.

Council of the District of Columbia. (2011). Distributed Generation Amendment Act of 2011, Permanent Legislation. http://www.dccouncil.washington.dc.us/images/00001/20110726120857.pdf.
Accessed November 2, 2011.

Database of State Incentives for Renewables and Efficiency (DSIRE). (2011a). "Rules, Regulations & Policies." http://dsireusa.org/incentives/index.cfm?EE=1&RE=1&SPV=0&ST=0&searchtype=RPS&sh=1. Accessed October 3, 2011.

DSIRE. (2011b). "Maryland Public Service Commission—Solar Renewable Energy Certificates (SRECs)." http://dsireusa.org/incentives/incentive.cfm?Incentive_Code=MD55F&re=1&ee=1. Accessed August 3, 2011.

DSIRE. (2011c). "Ohio SREC-Based Financing Programs." http://www.dsireusa.org/incentives/incentive.cfm?Incentive_Code=OH61F&re=1&ee=1. Accessed August 15, 2011.

DSIRE. (2011d). "New Jersey Board of Public Utilities—Solar Renewable Energy Certificates (SRECs)." http://dsireusa.org/incentives /incentive.cfm?Incentive Code=NJ07F&re=1&ee=1. Accessed August 3, 2011.

DSIRE. (2011e). "PSE&G – Solar Loan Program." http://dsireusa.org /incentives/incentive.cfm?Incentive_Code=NJ21F&re=1&ee=1. Accessed August 3, 2011.

Duke Energy Carolinas. (2011a). "Duke Energy Carolinas Standard Purchase Offer for Renewable Energy Certificates ('RECs')." http://www.duke-energy.com/pdfs/RECPurchase-Offer-Info.pdf. Accessed August 15, 2011.

Duke Energy Carolinas. (2011b). "Duke Energy's 2010 REPS Compliance Report and Application for Approval of REPS Cost Recovery." E-7 Sub 984. http://ncuc.commerce.state.nc.us/cgibin/fldrdocs.ndm/INPUT? compdesc=Duke%20Energy%20Carolinas%2C%20LLC&numret =065&comptype=E&docknumb=7&suffix1=&subNumb=984&suffix2=&parm1=00013509 6. Accessed September 12, 2011.

Duke Energy Carolinas. (2011c). "Docket No. E-7, Sub 856, 2010 Annual Cost Update." http://ncuc.commerce.state.nc.us/cgibin/fldrdocs.ndm /INPUT?compdesc=Duke%20Energy%20Carolinas%2C%20LLC&numret =001&comptype=E&docknumb=7&suffix1=&subNumb =856&suffix2 =&parm1=00012889 6. Accessed September 12, 2011.

Duke Energy Ohio. (2011). "Solar Renewable Energy Credits Program." http://www.dukeenergy.com/ohio/renewable-energy/srec.asp. Accessed August 15, 2011.

Elefant, C.; Holt, E. (2011). *The Commerce Clause and Implications for State Renewable Portfolio Standard Programs*. Clean Energy States Alliance. http://www.cleanenergystates.org/resource-library/resource/cesa-report-the-commerceclause-and-implications-for-state-renewable-portfolio-standard-programs-pdf. Accessed August 15, 2011.

Epsen, K. (2 November 2011). Email. New Hampshire Public Utilities Commission, Dover, New Hampshire.

First Energy Corporation. (2011a). "FAQ—Rules." https://www.firstenergy corp.com/utilitypowerprocurements/pa/srec/files/Penn_Power_Docu ments/GENERAL FAQ- PP Solar RFP-2011-02-24.pdf. Accessed August 3, 2011.

First Energy Corporation. (2011b). "Pennsylvania Power Company Completes Purchase of Solar Renewable Energy Credits press release 3-21-11." https://www.firstenergycorp.com/utilitypowerprocurements/pa/srec/files/P enn_Power_Docu ments/2011-03-18 Pennsylvania Power Company Completes Purchase of .pdf. Accessed August 3, 2011.

FlettExchange. (2011). Settlement Price. http://www.flettexchange.com/. Accessed October 3, 2011.

Flynn, H.; Breger, D.; Belden, A.; Bier, A.; Laurent, C.; Andrews, N.; Rickerson, W. (2010). "System Dynamics Modeling of the Massachusetts SREC Market." *Journal of Sustainability* (2); pp. 2746–2761.

Hart, D.M. (2010). "Making, Breaking, and (Partially) Remaking Markets: State Regulation and Photovoltaic Electricity in New Jersey." *Energy Policy* (38); pp. 6662–6673.

Heeter, J.; Bird, L. (2011). *Status and Trends in U.S. Compliance and Voluntary Renewable Energy Certificate Markets (2010 Data).* Golden, CO: National Renewable Energy Laboratory.

Kansas City Power & Light (KCP&L). (2011). "2011 Annual Renewable Energy Standard Compliance Plan." Case No. EO-2011-0277. https://www.efis.psc.mo.gov/mpsc/. Accessed September 12, 2011.

Kansas City Power & Light General Missouri Operations (KCP&L-GMO). (2011). "2011 Annual Renewable Energy Standard Compliance Plan." Case No. EO-2011-0278. https://www.efis.psc.mo.gov/mpsc/. Accessed September 12, 2011.

Massachusetts Department of Energy Resources (Massachusetts DOER). (2011a). "Metering, Data Reporting, and Verification Requirements." http://www.mass.gov/?pageID=eoeeaterminal&L=5&L0=Home&L1=Ene rgy%2c+Utilities+ %26+Clean+Technologies&L2=Renewable+Energy &L3=Solar&L4=RPS+Solar+CarveOut&sid=Eoeea&b=terminalcontent &f=doer renewables solar metering-datareporting&csid=Eoeea. Accessed October 3, 2011.

Massachusetts DOER. (2011b). "RPS Solar Carve-Out Qualified Units." http://www.mass.gov/Eoeea/docs/doer/rps_aps/solar-carve-out-units.xls. Accessed November 1, 2011.

Massachusetts DOER. (2011c). "Current Status of the RPS Solar Carve-Out Program." http://www.mass.gov/?pageID=eoeeamodulechunk&L=5&L0=Home&L1=Energy%2c+Utilities+%26+Clean+Technologies&L2=Renewable+Energy&L3=Solar&L4=RPS+Solar+Carve-Out&sid=Eoeea&b=terminalcontent&f=doerrenewablessolarcurrentparameters&csid=Eoeea. Accessed September 30, 2011.

Massachusetts DOER. (2011d). "Solar Credit Clearinghouse Auction." http://www.mass.gov/?pageID=eoeeaterminal&L=5&L0=Home&L1=Energy%2c+Utilities+ %26+Clean+Technologies&L2=Renewable+Energy&L3=Solar&L4=RPS+Solar+CarveOut&sid=Eoeea&b=terminalcontent&f=doer renewables solar solar-credit&csid=Eoeea. Accessed August 3, 2011.

Moore, G. (2011). "Utility Collaboration in Delaware's SREC Market." Presentation by Delmarva Power to Solar Power International, October 28, 2011.

Mosier, K. (10 August 2011). Personal communication. Maryland Public Service Commission, Baltimore, MD.

NERA Economic Consulting. (2011a). "NJ EDC Solar—Program Guide (July 18, 2011)." http://njedcsolar.com/documents.cfm. Accessed August 3, 2011.

NERA Economic Consulting. (2011b). "SREC-Based Financing Program Update." Available at: http://njedcsolar.com/assets/files/SREC-Based_Financing_Program_Update_7-28-111.pdf Accessed November 6, 2011.

New Hampshire Public Utilities Commission (NH PUC). (2011). "2010 Annual Compliance Report." http://www.puc.state.nh.us/Sustainable%20Energy/Renewable_Portfolio_Standard_Program. htm. Accessed November 2, 2011.

New Jersey Board of Public Utilities (NJ BPU). (2008). "NJ BPU Solar Financing Board Order." http://njcleanenergy.com/files/file/Board%20Orders/7-30-08-8E.pdf. Accessed August 3, 2011.

NJ BPU. (2011a). "Board Order Approving Results of the Sixth Solicitation (March 30, 2011)." http://njedcsolar.com/documents.cfm. Accessed August 3, 2011.

NJ BPU. (2011b). "Board Order Approving Results of the Seventh Solicitation (July 14, 2011)." http://njedcsolar.com/documents.cfm. Accessed August 3, 2011.

NJ BPU. (2011c). "SREC Registration Program Status Reports." http://www.njcleanenergy.com/renewable-energy/program-activity-reports/programstatusreports/srec-registration-program-status-reports. Accessed September 30, 2011.

NJ BPU. (2011d). "New Jersey's Renewable Portfolio Standard Rules: 2010 Annual Report." April 13, 2011 version. Available at: www.njcleanenergy.com/files/file/RenewablePrograms/Draft 2010 Annual Report for New_Jersey_041311_version.pdf Accessed November 6, 2011.

NJ BPU. (2010). "Board Order Approving Results of the Fourth Solicitation (August 12, 2010)." http://njedcsolar.com/documents.cfm. Accessed August 3, 2011.

North Carolina Green Power (NCGP). (2011). "Become a Renewable Energy Generator." http://www.ncgreenpower.org/resources/. Accessed October 4, 2011.

North Carolina Utilities Commission (NCUC). (2011a). "NEW Renewable Energy Facility Registrations 2008–2011." www.ncuc.commerce.state.nc.us/reps/RegistrationSpreadsheet2008-2011.xls. Accessed November 1, 2011.

NCUC. (2011b). "Duke Energy Carolinas Application for Approval of Renewable Energy and Energy Efficiency Portfolio Standard Cost Recovery Rider, Docket E-7, Sub 984. Transcript of Testimony—Volume 1 (heard June 8, 2011), Cross-Examination by Mr. Olson of Kim Smith and Emily Felt." p. 87. http://ncuc.commerce.state.nc.us/cgibin/webview/senddoc.pgm?dispfmt=&itype=Q&authorization=&parm2=2BAAAA76111B& parm3=000135096. Accessed September 12, 2011.

Ohio Edison Company; Cleveland Electric Illuminating Company; Toledo Edison Company. (2010). "Annual Status Report and 2009 Compliance Review. Case No. 10-499-EL-ACP." http://dis.puc.state.oh.us/TiffToPDf/A1001001A10D15B51903F44369.pdf. Accessed October 3, 2011.

Pennsylvania PUC. (2011). "Final Policy Statement Order—Public Meeting of Sept. 16, 2010." http://www.puc.state.pa.us/electric/electric alt energy.aspx. Accessed August 3, 2011.

PJM-Generation Attribute Tracking System (PJM-GATS). (2011a). "RPS Retired Certificates (Reporting Year)." https://gats.pjmeis.com/myModule/rpt/myrpt.asp?r=241&TabName=All. Accessed November 1, 2011.

PJM-GATS. (2011b). "Renewable Generators Registered in GATS." https://gats.pjmeis.com/myModule/rpt/myrpt.asp?r=228. Accessed October 31, 2011.

PJM-GATS. (2011c). "Solar Weighted Average Price." https://gats.pjmeis.com/myModule/rpt/myrpt.asp?r=230. Accessed October 3, 2011.

PPL Electric. (2011a). "Provider of Last Resort, AEPS. Solar Set-Aside Product." http://www.pplelectric.com/Business+Partners/polr-aeps-solar+set-aside+product/. Accessed August 3, 2011.

PPL Electric. (2011b). "Results of May 2011 Long Term Solar Solicitation." http://www.pplelectric.com/NR/rdonlyres/8A3D20E8-2BCC-4795-8992-1A24B57B872C/0/PAPUC SecretarialLetter 17052011.pdf. Accessed August 3, 2011.

Progress Energy Carolinas. (2011a). "Application for Approval of REPS Cost Recovery Rider and Direct Testimony of Jay Foster." E-2 Sub 1000. http://ncuc.commerce.state.nc.us/cgibin/fldrdocs.ndm/INPUT?compdesc=PROGRESS%20ENERGY%20CAROLINAS%2C%20INC.%3B%20CAROLINA%20POWER%20%26%20LIGHT%20COMPANY%2C%20DBA&numret=065&comptype=E&docknumb=2&suffix1=&subNumb=1000&suffix2=&parm1 =000135584. Accessed September 12, 2011.

Progress Energy Carolinas. (2011b). "Sunsense Solar PV Program." https://www.progress-energy.com/carolinas/business/save-energy-money/sunsense/commercial-solar-pv-program.page. Accessed August 15, 2011.

Public Service Commission of Maryland (PSC MD). (2011). "Renewable Energy Portfolio Standard Report of 2011, With Data for Compliance Year 2009." http://webapp.psc.state.md.us/intranet/Reports/MD %20 RPS%202011%20Annual%20Report. pdf. Accessed October 4, 2011.

Sherwood, L. (2011). "U.S. Solar Market Trends 2010." Latham, New York: Interstate Renewable Energy Council. http://irecusa.org/wp-content/uploads/2011/07/IREC-SolarMarket-Trends-Report-revised070811.pdf. Accessed September 30, 2011.

Siegfried, S. (15 July 2011). Personal communication. Ohio Public Utilities Commission, Columbus, OH.

Solar Electric Power Association (SEPA). (2011). "Solar Data and Mapping Tool." http://www.solarelectricpower.org/solar-tools/solar-data-and-map ping -tool.aspx. Accessed October 4, 2011.

Solar Energy Industries Association (SEIA). (2011). "Utility-Scale Solar Projects in the United States Operating, Under Construction, or Under Development." www.seia.org/galleries/pdf/Major%20Solar%20Projects.pdf. Accessed September 30, 2011.

SEIA and GTM Research (GTM). (2011a). "U.S. Solar Market Insight: 2010 Year-InReview."

SEIA and GTM. (2011b). "U.S. Solar Market Insight: 2nd Quarter 2011."
SEIA and GTM. (2011c). "U.S. Solar Market Insight: 1st Quarter 2011."
SRECTrade. (2011a). "State SREC Markets." http://www.srectrade.com /background.php. Accessed October 3 2011.
SRECTrade. (2011b). "50 MW Ohio Solar Project Secures Further Financing." http://www.srectrade.com/blog/srec-markets/50-mw-ohio-solar-project-secures-furtherfinancing. Accessed October 3, 2011.
SRECTrade. (2011c). "SREC Market Prices." http://srectrade.com/srec prices.php. Accessed October 4, 2011.
SRECTrade. (2011d). "New Jersey Installed Capacity Update." http://www.srectrade.com/blog/srec-markets/new-jersey-installed-capacity-update. Accessed October 3, 2011.
SRECTrade. (2011e). "Pennsylvania Legislature Commences Efforts to Fix SREC Program." http://www.srectrade.com/blog/srec-markets /pennsylvania/pennsylvania-legislaturecommence-efforts-to-fix-srec-program. Accessed October 24, 2011.
Stewart, C. (22 July 2011). Personal communication. Delaware Public Service Commission, Dover, Delaware.
Tennessee Valley Authority (TVA). (2011). Generation Partners. http://www.tva.com/greenpowerswitch/partners/index.htm. Accessed October 4, 2011.
Union of Concerned Scientists (UCS). (2008). "Summary Information of State RPS Programs." www.ucsusa.org/assets/documents/clean energy/pennsylvania.pdf. Accessed September 30, 2011.
Wiser, R.; Barbose, G.; Holt, E. (October 2010). *Supporting Solar Power in Renewable Portfolio Standards: Experience from the United States*. LBNL-3984E. Berkeley, CA: Lawrence Berkeley National Laboratory.

APPENDIX

Long-Term Contracting Provisions: SREC Auctions in New Jersey

In 2008, the New Jersey Board of Public Utilities (the "Board") established a three-year, long-term contracting program for SRECs for the state's regulated distribution utilities (NJBPU 2008). The program is called the SREC-Based Financing Program because these auctions are intended to help winning residential and commercial developers secure financing for their solar projects. The contracts resulting from the auctions are standardized and non-negotiable, and developers are encouraged to bid for SREC contracts that would provide sufficient project returns. The Board has included a provision that precludes a project that receives an award in one solicitation to compete in a future solicitation at a higher price absent a "clear indication of a change in circumstances" (NJ BPU 2010).

For RPS Reporting Years 2010–2012, Jersey Central Power and Light Company, Atlantic City Electric Company, and Rockland Electric Company are required to hold at least two auctions annually to select competitive proposals for SRECs under long-term contracts. To recover associated costs, the utilities are able to auction the SRECs procured in the SRECBased Financing Program to the state's generation suppliers.

The auctions are held for SREC contracts between 10 and 15 years. This length enables solar projects to obtain financing and addresses a barrier to development. The bidder submits a pricing proposal ($/SREC and contract duration) and a summary of project qualifications (NERA Economic Consulting 2011a). Since bids may vary in terms, the auctioneer ranks each bid according to the net present value of payments under the SREC agreement. There is an undisclosed price limit when evaluating bids, and the auctioneer decides whether or not there are enough competitive bids to meet the auctioned quantity (NERA Economic Consulting 2011a).

The Board-mandated program has the goal of procuring more than 64 MW over three years: 42 MW for Jersey Central Power and Light Company, 19 MW for Atlantic City Electric Company, and 3.769 MW for Rockland Electric Company (NERA Economic Consulting 2011b). For each solicitation, the Board publishes a planned quantity/capacity cap for each EDC, and the auctioneer accepts bids in order of competitiveness up to that cap. If in a given solicitation the planned quantity is not procured, the Board assigns the shortfall to subsequent auctions. In RPS reporting years 2010 and 2011, the three EDCs procured a total of 47.5 MW in six separate solicitations. The remaining 16.5 MW of the total program target will be procured in two or three solicitations during the 2012 RPS reporting year (NERA Economic Consulting 2011b).

In 2010, the program expanded eligibility to residential and commercial programs up to 2 MWDC in size, from the earlier 500 kW limit, to increase participation (NERA Economic Consulting 2011a). The program administrators consider three segments of project sizes: a small project size segment (<50 kW); a mid-size project segment (>50 kW, <500 kW); and a large project size segment (>500 kW, <2 MW).

Long-Term Contracting Provisions: Competitive Solicitations in Pennsylvania

In September 2010, the Pennsylvania PUC adopted a "Policy Statement in Support of Pennsylvania Solar Projects" (see PUC Code §69.2901-2903). The policy statement encourages EDCs to issue RFPs for large-scale solar projects (>200 kW) using a transparent and open competitive bidding process. The statement also suggests that EDCs can procure SRECs from small-scale projects (<200 kW) through both RFPs and bilateral contracts. The policy statement specifies that the price negotiated for SRECs from small-scale projects should not exceed the commission-approved average winning bid price in the EDC's most recent RFP for large-scale solar projects (Pennsylvania PUC 2011).

In response, Pennsylvania EDCs have started to hold competitive solicitations to procure a portion of their SREC obligations over a longer delivery period (e.g., 7–10 years). Four of Pennsylvania's 11 EDCs have held RFPs for long-term SREC contracts to date, including: Metropolitan Edison Company, Pennsylvania Electric Company, Pennsylvania Power Company, and PPL Electric Utilities.

The utilities have structured their procurements so that long-term contracts expire in 8–10 years (2020) when the Pennsylvania alternative energy portfolio standard concludes. The utilities are seeking bids in tranches (i.e., 100, 250, or 500 SRECs per year from an individual supplier). SREC suppliers have the responsibility of delivering the quantity of SRECs specified by their contract each year. One advantage of this approach is that the utility does not have to bear the risks associated with individual project performance (Bingaman 2010).

SREC Price Floor Programs: Massachusetts and New Jersey Utility PSE&G *Massachusetts Solar Credit Clearinghouse*

In 2010, Massachusetts created the Solar Credit Clearinghouse Auction, which functions as a long-term price floor for SRECs in the Massachusetts market. Massachusetts opted for this approach because there was little appetite to adopt long-term contracting provisions at the time the solar carve out was implemented (Bernstein et al. 2011).

Under the clearinghouse program, the Massachusetts Department of Energy Resources (DOER) will hold an annual fixed-price SREC auction where generators can deposit eligible, unsold SRECs into an auction account with the NEPOOL-GIS. Solar generators may deposit SRECs into the account during the final 31 days of each compliance year's Q4 trading period (May 16–June 15). If any SRECs are deposited, DOER must hold an auction to clear these SRECs prior to July 31 of that year. Bidders submit the volume that they are willing to buy for the fixed price of $300/MWh, the floor price for the SRECs (Massachusetts DOER 2011d). There is a 5% auction fee, so generators will receive $285 for each SREC cleared on the auction (instead of $300/SREC).

At the time of qualification, generators will be given an "opt-in term," which signifies the number of quarters that their project will have the opportunity to deposit SRECs into the auction account. The opt-in term for new projects is currently 10 years. The opt-in term for future projects may be adjusted annually based on market conditions, but any adjustment will have no impact on projects that have already been qualified. If the SREC market is oversupplied, the DOER will decrease the opt-in period for new systems to temper market growth (Flynn et al. 2010). For every 10% of the compliance obligation met by deposits into the auction, the opt-in term will be reduced by one year. It will not decrease more than two years at a time. In the event of a shortage, the opt-in term will increase by one year for each 10% of the compliance obligation met through solar ACP payments. The opt-in term may never be longer than 10 years. It cannot decrease below five years until after 2016.

When SRECs are deposited into the auction account, they are re-minted with an initial shelf life of two years, meaning that they can be used for compliance for either one of the next two compliance years. The compliance obligation the following year is increased by the amount of SRECs deposited in the auction. If there is insufficient bid volume to clear the auction account SRECs, the shelf life is increased to three years, and the auction is repeated (Massachusetts DOER 2011d). If the second auction does not clear, a third and final auction will be held. SRECs retain the three-year shelf-life from the second auction, and the compliance obligation for the following year is increased a second time by the amount of SRECs in the auction account. If there are insufficient bids to clear this final auction, depositors will be paid on a pro-rated basis for the number of SRECs deposited relative to the number sold. Any remaining SRECs will be returned to depositors, who will have three years to sell them.

The auction is intended as a market of last resort. Once a SREC is moved into the auction account, it cannot be removed (Massachusetts DOER 2011d).

The auction mechanism has not yet been used because there has been a shortage of SRECs to date.

PSE&G Solar Loan Program

The Public Service Electric & Gas Company (PSE&G) Solar Loan Program has elements of a long-term price floor. Under the program, PSE&G lends money to its retail customers to cover 40%–60% of the cost of a PV system. Customers may repay the loan principal and interest through cash payments or by signing their SRECs over to PSE&G. The loan term is 15 years, and during this period, PSE&G credits customers for each SREC generated according to market conditions but not below the established price floor. The price floor as of January 2011 is $420/SREC for residential borrowers and $380/SREC for commercial borrowers. The price floor declines over time; however, the floor price applicable at the time of loan closing remains in effect for a particular project for the entire duration of the loan (DSIRE 2011e). Customers will be paid at least the floor price for each SREC generated, but if trading prices are higher, customers may be paid more. PSE&G reserves the right to purchase SRECs generated by the system at 75% of the market price through the original term of the loan (DSIRE 2011e). The Solar Loan I program, which was initiated in 2008, had the goal of installing 30 MW of customer-sited PV. Solar Loan II, which began in December 2009, has a current program target of 51 MW of installed capacity (NJ BPU 2011d).

End Notes

[1] In states that have undergone electricity market restructuring, electricity customers have retail choice, meaning that they can shop among electricity suppliers. EDCs own the distribution infrastructure and are the providers of last resort. EGSs are suppliers that have entered the market to provide electricity generation service.

[2] Publicly owned utilities are obligated in Maryland, while they are encouraged to comply voluntarily in Pennsylvania (UCS 2008) and can opt out of the Delaware RPS and solar set-aside if they create their own RPS standard by 2013. In North Carolina, a traditionally regulated electric market, investor-owned utilities (IOUs), municipal utilities, and electric cooperatives are required to meet the solar electricity requirement, though their overall RPS targets vary (12.5% for IOUs and 10% for munis/co-ops).

[3] Solar thermal is also eligible to meet solar set-aside targets in states that do not use SRECs. These states include Arizona, Nevada, and New York (Wiser et al. 2010).

[4] As of June 30, 2011, more than 940 MW of PV was installed or actively reserved through the California Solar Initiative program (CPUC 2011).

[5] Other mechanisms have been used to minimize or address investor risk, but these are generally outside the scope of SREC markets. Also, some states have used centralized procurement for their RPS.

[6] PJM-Generation Attribute System (PJM-GATS), the REC tracking system for the PJM balancing authority, provides information about the source of SRECs retired for each solar set-aside in the region. The majority of states with SREC markets are in PJM. This information is available on the Public Reports page of the PJMGATS website, under the heading "RPS Retired Certificates (Reporting Year)" (PJM-GATS 2011a). Massachusetts, Missouri, New Hampshire, and North Carolina employ different systems to track RPS compliance. Missouri uses the North American Renewables Registry (NARR), Massachusetts and New Hampshire use the NEPOOL Generation Information System (GIS) in New England, and North Carolina uses the NC-Renewable Energy Tracking System (NCRETS), which currently accepts credit transfers from the Midwest Renewable Energy Tracking System (M-RETS), NARR, the Western Renewable Energy Generation Information System (WREGIS), and the Electric Reliability Council of Texas (ERCOT) REC tracking system.

[7] PV systems in North Carolina must choose whether to register in NC-RETS or the PJM-GATS system.

[8] Typically projects are certified by state agencies for compliance and then registered in PJM-GATS. Registered generators are all generating units that have an account with the GATS administrator. The facilities in the GATS database are operational.

[9] No PV systems are registered yet in Missouri's chosen tracking system, NARR. The first compliance year in Missouri is 2011.

[10] Under the Duke Energy PV DG Program, Duke has secured 25 sites and installed roughly 9.9 MWDC (estimated to generate 13,485 MWh annually) (Duke Energy 2011c).

[11] New Jersey has no specific limits, but systems must be connected to the distribution grid, and behind-themeter projects must be sized so that annual output does not exceed annual on-site load (DSIRE 2011d).

[12] PJM-GATS reports solar weighted average prices for transactions in the PJM market (see PJM-GATS 2011c) that differ from spot prices reported by SRECTrade and Flett Exchange because PJM-GATS pricing can include pricing from long- or mid-term contracts as well as spot prices. PJM-GATS reports prices on a monthly basis, based on when the SREC was issued, traded, or retired, not on when the generation occurred (PJM-GATS 2011c). Recently, due the decline in spot pricing, PJM-GATS has reported higher solar weighted average prices than the spot prices reported by SRECTrade and Flett Exchange. For example, if a company contracted for SRECs that were generated in January 2010 at a given price but did not retire those SRECs until August of 2011, the January 2010 price would be reflected in PJM-GATS's August 2011 solar weighted average price report.

[13] These figures do not include completed projects.

INDEX

H

I

J

L

M

N

O

P

Q

R

S

T

U